Ice, Rock, and Beauty

David Brodie

Ice, Rock, and Beauty

A Visual Tour of the New Solar System

Project Adviser

Professor Carl Murray,
Queen Mary, University of London,
and Member of the Cassini Imaging Team

 Springer

David Brodie
hiphop@dircon.co.uk

ISBN 978-0-387-73102-5   e-ISBN 978-0-387-73103-2

Library of Congress Control Number: 2007931123

Acknowledgments

This is a book for inhabitants of the Solar System, or at least for those who take an interest in this neighborhood of ours. Recognition and deepest thanks are due to all of those involved in originating the wonderful pictures of objects near, far, and very far. I have made every effort to contact image copyright holders and to provide accurate credit lines. If there are any errors or omissions then I shall pleased to correct them at the first opportunity.

I am grateful to Carl Murray for his comment on the accuracy of information, and to Rosemary Phillips for her proof reading support and general patience. Any surviving failings are entirely my responsibility.

Aberystwyth, August 2007   David Brodie

Contents

Introduction

The Revolutions of Crystal Spheres

A lot can happen in 400 years – very many performances of Shakespeare's Hamlet, for example, since the first one early in the 1600s; and four hundred orbits of the Sun for the Earth and its transient passengers.

During earlier orbits, passengers who knew and cared enough were generally of the view that the Sun's motion consisted of revolutions around the static Earth. Such was their faith in their own importance they were convinced that the Earth was the center of all that existed. At the same time, they knew it as a place of corruption, rough-edged and rough, a platform for lives that were so often short and brutish. Surely there had to be something better, and if not here then in the heavens. There, they asserted, were the realms of perfection, of crystal spheres carrying the pure lights of constant stars and of roaming planets.

At the start of 1610, observers aided by the latest in optical technology, the telescope, saw some small "stars" near Jupiter. It is likely that at least some of them had been seen before, with the sharp eyes of Chinese and Islamic observers, for example, but never in enough detail for another look on the following night to show that these stars move, relative to each other and to Jupiter.

Stars are not supposed to do that. Their patterns stay fixed from one millennium to the next, while it is the planets like Jupiter that wander among them. These "new" points of light were seen to travel with Jupiter, loosely, forming different patterns as each night went by.

There was, in the end, only one explanation. They were companions of Jupiter, small moons in orbit around the giant planet. These moons were an imperfection in the heavens; they did not orbit the Earth but smashed through Jupiter's crystal sphere and thus added to the growing doubt about the Earth-centered model of the Universe.

The Solar System lost its perfection. Little by little it became ever more rough-edged. The mountains of the Moon destroyed the status of our bright neighbor as a perfect sphere – perfection was a quality imposed upon it by humans, and an expectation that it could not match. Sunspots observed from day to day suggested that the Sun not only rotated around an internal axis, but had a blighted face. Saturn, seen from Earth, was not even round at all, but accompanied by huge discs. The surface of Mars had features that eventually came to be recognized as mountains, cliffs, and ravines.

Observation overruled ancient dogma, and the Sun became the center of motion of the wandering planets. The revolutions of the skies were turned around. The revolutions in peoples' minds were traumatic. Most shocking of all was the new knowledge that our own motion was around the Sun – our own home was one planet among several.

The moons of Jupiter may be a long way away. They may seem very small. But they are the "stars" that heralded the birth of science. They helped to change our approach to the power of our own observations in the face of old authority. Thus they changed the way we think and the way we live.

The Beauty of Imperfect Worlds

Now, with space missions that stream images back to us, we can see that the Solar System is full of bodies whose imperfection is what makes them so wonderful. Apart from the moons of Jupiter, the mountains of the Moon, the rings of Saturn, we can see in detail almost as if we could be there, the fiery pits of Io, the colored mists of Titan, the crystal fountains of Enceladus.

Jupiter and three of its moons
Credit: NASA/JPL/Malin Space Science Systems

Beauty is indeed in the eye of the beholder who imposes his or her responses back onto objects that just *are*. The bodies of the Solar System, and the Earth itself, have no subjective qualities except inside our heads. Beauty is such a quality, and perhaps one that gives us inner protection from the huge indifference of a world that just is.

Though it might be optimism to the point of absurdity, it is not hard to experience beauty. There is no need to seek the idealized beauty of the ancients when imperfect beauty is so easily found; and so vast. It starts right where we are and goes on, through the clouds and above them until the blue fades to black, beyond that to the Moon, to every planet and every moon, to the quiet hordes of asteroids, to comets with their misty tails, and beyond to the stars. The Earth's wonderful imperfection is not unique; there are so many other imperfect worlds, so much the same and so much different.

The New Solar System

A lot has, indeed, happened in the last 400 years. The planets are no longer points of light that wander across the skies. We have seen them through telescopes and we have sent our robotic image-gatherers, such as Voyagers 1 and 2, Galileo and Cassini, to take a look up-close.

Recent schoolbooks still talk of nine planets and not all that many more moons. Discoveries of so many small moons, asteroids, and countless distant bodies have changed everything.

We are looking at a new Solar System of planets and of smaller bodies. The images in this book have been chosen for the stories they tell about its depth and diversity, and for their invitations to new experiences of imperfect beauty

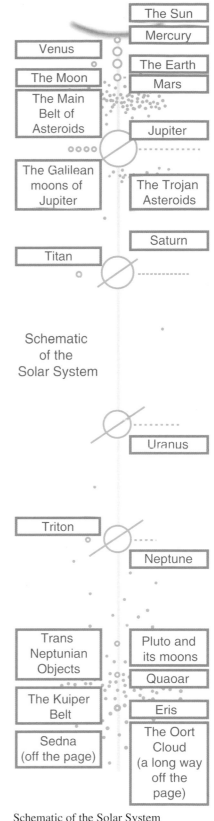

Schematic of the Solar System

The Inner Solar System

Everything in the Solar System, from the tiniest grain to the giant Jupiter, is a child or a prisoner of the Sun. Each one stays in orbit, in preference to escape to the cold freedom of interstellar space. Some stay closer, much closer, than others.

Mercury, the innermost planetary resident, is moonless and indeed smaller than Jupiter's Ganymede or Saturn's Titan. As a conventional planet and not one of the Solar System's multitude of bodies of more recent discovery, it features no further in this book.

Venus likewise follows a solitary track around the Sun. One of the long-running scientific controversies of the past concerns a "moon" that many observers claimed to see. It was even named, Neith, but continuing observations showed that Venus had no such close companion. The planet itself is a deadly place. Part of its history is a runaway greenhouse effect – an accumulation of carbon dioxide has raised the temperature of surface and atmosphere to levels that will melt lead – and part of the present is its global acid mist. Venus is the best warning we have of just how bad things can get.

Among these various bodies of the inner Solar System are the Earth and its Moon. They are partners; each one softens the night of the other. In times of greater everyday (or every-night) sky literacy than our own, moonlight was a precious aid to living. That living, our own existence among the complexity of the planet's life, is itself a marvel of the Solar System, and it does not lessen that marvel to have discovered complexity in many and various guises in worlds, large and small, beyond our own.

Despite the ancient stories that link Venus with love and Mars with hate, the latter is a friendlier place, or at least it is not quite so hideously inhospitable. It's just a global desert, that's all.

It is quite likely that Mars would be alone in its travels, like Venus, if it were not for the relative proximity of the Main Belt of asteroids, a ring of rocks around the Sun. Two suspiciously asteroid-like objects, Phobos and Deimos, fly over the surface of Mars, captured by its gravity, and the planet has other small companions that share its orbit.

Just as the non-existent moon of Venus had a name, so is there a name for a planet that probably never existed – Phaeton. One suggestion for the origin of the Main Belt of asteroids, between the orbits of Mars and Jupiter, is that they are the remains of this destroyed planet. However, the total mass of all the bodies in the belt is a lot less than that of the Moon. It seems more likely that the process by which matter pulls on matter to form planets was here opposed by the influence of Jupiter's gravity and the mutual jostling of the grains and rocks, so that planet-building was just not possible.

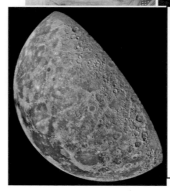

Contrasts

One of the lesser planets travels with one of the larger moons. The smaller body of the two is inert, geologically, magnetically, and biologically. But the planet has a stable continuity – conditions that keep on changing, sustaining life through it all, and giving rise to consciousness, observation, record-making, words and images, and books like this. This planet, like many of the bodies of the Solar System, is not an inert place.

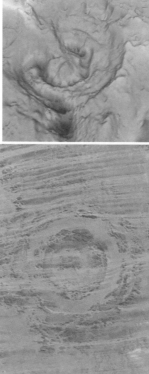

Changing, healing

The ragged edges of the Moon are clear through a telescope. It is a body pitted with craters – existence is impossible, unshielded in a spinning disk of rocks of unknown number, without gaining scars. Despite its protective blanket of air, the Earth has not escaped a battering, but its dynamism has healed old wounds and only the most recent have survived to be witnessed by ourselves.

The battering goes on

The bombardment of the Earth continues through every 24 hour axis-spin, all 365 or so of them in each solar orbit. The quick streaks of light across the night sky are not hard to see, with just a little patience. They are meteors. Occasionally one lights up the sky, brighter to us than any star, for just a brief flicker. And sometimes they are big enough to survive their journey through the air and reach the surface, on land or sea, and we can find them – bodies of stone or metal or a mixture – meteorites. There are larger bodies out there, asteroids and comets, with the potential to crash down upon us, and certainly the dinosaurs suffered. Fortunately, such megacollisions don't happen often.

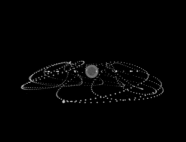

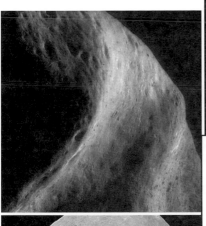

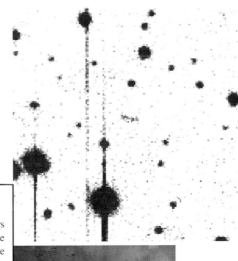

Assorted rocks

The asteroids of the "Main Belt" between Mars and Jupiter are not, in fact, all "rock" – some are metallic or stone–metal mixtures. They are assorted in size and orbital behavior as well as composition. Many asteroids are mere pebbles and grains. A small proportion of them are more than 100 kilometers across, and just one, Ceres, is big enough to be called a dwarf planet. Their distances from the Sun vary between roughly twice and four times our own orbital radius. It seems that some were subject to Jupiter's gravity so that they were thrown out of more regular orbits long ago, creating gaps and clusters in what would otherwise have been a more evenly distributed belt of asteroids. The small moons and other companions of Mars are almost certainly asteroids that have strayed from their original orbital homes.

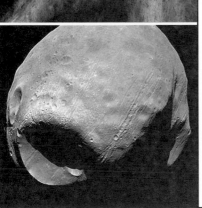

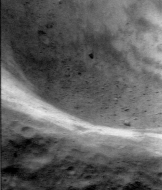

Followers of Jupiter

Jupiter (alias Jove) may not look like anything much, at least at a casual glance, other than just another point of light in the sky. But even binoculars are enough to show its moons and its stripes. There is a lot going on up there.

There are storms in the deep, deep gassy atmosphere. It has volcanoes and crusts of cracked ice on its moons. There are auroras at its poles, and faint rings around it.

It has affected us powerfully.

Physically, its gravity protects us from greater bombardment by the smaller bodies of the Solar System. It has swept up many a comet and asteroid that otherwise might have created havoc by now on our planet. Though human lifespans are short compared with the long time through which the planets have circled, we have witnessed and recorded one such comet collision with the giant planet.

Culturally, it helped to eject European thinkers from their self-appointed, self-important place at the center of the Universe. Or rather, the discovery of its moons, Io, Europa, Ganymede, and Callisto did so, 400 hundred years ago. It helped to convince many that observation has greater power than ancient dogma in revealing truth about the physical world. It helped to create science which, along with associated technology and the ongoing mass escape from poverty, is a defining feature of our age and culture.

More recently, robot spacecraft that we have sent to do the observing for us have revealed new detail of Jupiter's weather systems and of the surfaces of the moons.

Jupiter does not do moons by halves. About 40 of them have human-given names, and there are plenty more, mostly quite small, so far untouched by our obsession for labels. All of these planetary followers are tiny compared with giant Jupiter, each one like a little droplet splashed from a bucket.

Moons are not all. There are also Trojan asteroids – two huge armies of ice or rock, one leading Jupiter in its orbit and the other following. The number of asteroids in these two groups is as large as that in the Main Belt, much further in toward the Sun. They are a long way away and no spacecraft has approached them, and so there are no exciting close-up images. But they are out there, just vague traces in the telescope images of astronomers – expert observers who know what they are looking for.

There is much about our own planet that we do not understand. There is much that we cannot predict. Comparison of Jovian phenomena and those on Earth, such as the weather in our atmosphere or the behavior of our rocky crust, provides scientists with opportunities to develop new techniques of understanding and predicting, new mathematical models and new visual interpretations. Whatever we can learn from Jupiter and its companions could be valuable.

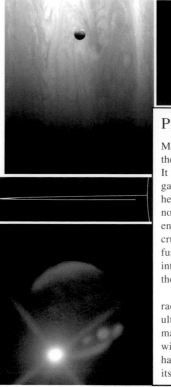

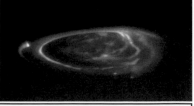

Planetary phenomena

More than 70% of all of the mass of the Sun's planets is carried by Jupiter. It is a giant of a planet, almost a star; a gassy ball made mostly of hydrogen and helium, much like the Sun. But it does not shine like the Sun. It is just not big enough to create the intense gravitational crush at its heart that can trigger nuclear fusion and a brilliant flow of energy out into its surroundings. It compensates, though, in other ways.

It emits radio waves and thermal radiation, its polar auroras send out ultraviolet light, and it reflects sunlight magnificently. It has a colored surface with characteristic bands of weather, has satellites of great diversity, and has its subtle rings.

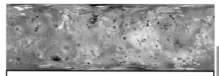

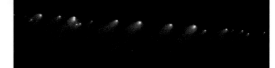

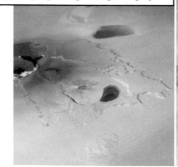

Interactive Io

The ongoing interaction of Io with its parent planet, under the influence of its sister moons, pumps its surface layers in and out, creating the moon's volcanoes. They release material that tumbles at speed toward Jupiter, becoming ionized by the violence of it all, and triggering Jupiter's polar light displays.

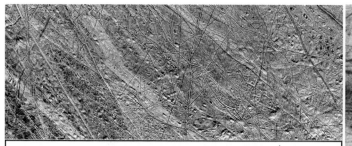

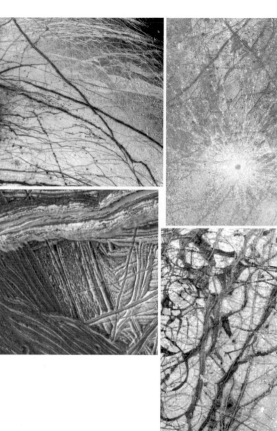

Europa

Io is dramatic and hot headed. Europa has cool and quiet beauty. It is about the size of our own Moon, but unlike our Moon it has an atmosphere though only a very, very sparse one. And it is an atmosphere with oxygen, created by decomposition of water molecules. Beneath this thin "air," processes of slow change in the icy crust have resulted in some wondrous features.

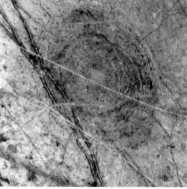

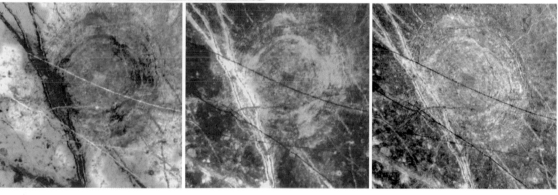

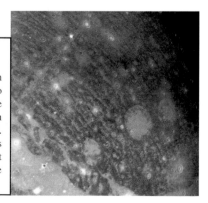

Ganymede and Callisto

These two moons, first and third in size in the Solar System, third and fourth in distance from Jupiter of the planet's four large moons, might be expected to be similar. They are not. Callisto seems a solid body, made of a mixture of ice, rock, and iron, similar to Titan and Triton. Ganymede has a layered structure, with molten iron in the core, creating a magnetic field that makes an interesting distortion in that of Jupiter. The surfaces of the two moons are also very different. Callisto is heavily cratered with otherwise little or no evidence of movement in the surface layers themselves – this is a moon of little change. Ganymede has craters, too, but also dark and light areas, grooves, mountains – all signs of intense geological activity, even though it may have happened long ago.

Saturn's Glories

Every great figure has followers, and Saturn has a horde of them. Courtiers are usually lesser bodies that cling to their leader, basking in reflected glory, but many of Saturn's followers have glory enough of their own.

There are, for a start, those rings. Early observers with their telescopes described them as being like "handles," and were nonplussed about what they might be. They could not dream that one day a robot spacecraft would beam close-up images back to Earth. But that is what the Cassini mission has done, providing pictures of many and various moons and of the grainy rings and their interactions with the small satellites that move almost among them.

Of the moons, many were known and named in the seventeenth century, in classical tradition: Titan, Iapetus, Rhea, Tethys, Dione. Others were not discovered until the Voyager 1 spacecraft approached in 1980: Atlas, Prometheus, Pandora, and more. For some, the first contact with human consciousness did not come until Cassini acted as the go-between, such as for Pallene in 2004 and Daphnis in 2005. (There has been a partial cultural shift of late, and some recent discoveries have names from Inuit, Norse, and Gaulish myths.)

The Huygens lander piggy-backed on Cassini all the way to the Saturnian system, and fell to the surface of Titan early in 2005. It found a little world that reminds us of our own. But appearances can be deceptive. Titan is an interesting place, certainly, and an object from which we can learn a great deal about moons and planets in general. It has much to teach us about the Earth in particular, though from Earth–Titan differences as much as from their similarities.

Saturn is indeed a great figure, but it is an airy object compared with its followers that are made of ice and rock. It is a gas giant, and the least dense of the four such bodies in the Solar System, less dense than water. Its small rocky core is surrounded by a layer of metallic hydrogen, but upward from there it is all gas.

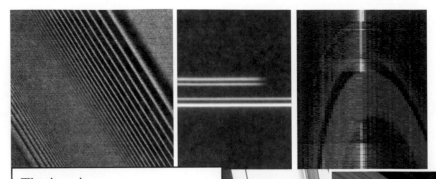

The icy rings…

The rings are brightly reflective, and form a bold visual feature; they appear huge but if their material were combined into a single body it would be a moon of small proportions. Their structure is very thin – wafer-like compared with their widths that stretch outward for almost half a million kilometers above the surface of the planet.

Their brightness is significant. Very old bodies would have gathered dust darkened by solar radiation, so it seems that the rings must be very much younger than the Solar System itself. There are suggestions about how they formed, perhaps by the break-up of a moon or comet, but so far there is no winning theory. Scientists are cautious as well as argumentative and more evidence will be needed before a single theory can win broad acceptance.

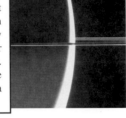

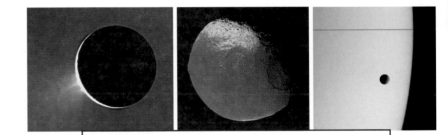

…and icy moons

The order of discovery of Saturn's mid-sized icy moons approximates to their order of size – Rhea, Iapetus, Dione, Tethys, Enceladus, Mimas. They form a family group, sisters perhaps, but it is a family that has grown up and grown apart. Events have shaped them to the moons they have become, and they have each taken on their own outward characteristics. Most exciting among them is Enceladus, with its internal heat source driving its ice fountains that in turn feed material into one of Saturn's rings.

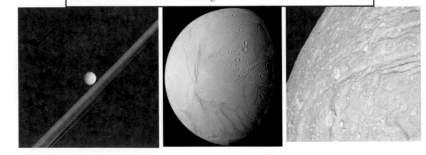

Oddballs and origins

Hyperion and Phoebe are not part of the same family as Saturn's larger moons. Phoebe has every appearance of a captured object, a comet nucleus, with origins in the outer Solar System. Hyperion has every appearance of something very strange indeed, whose history can, for now, only be the subject of hypothesis.

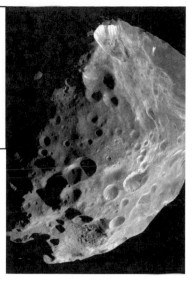

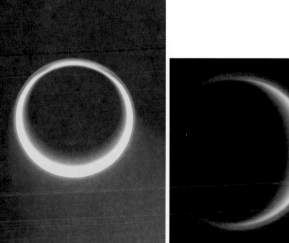

Neither Venus nor Titan

Just as people once dreamt that our neighbor Venus could provide us with an alternative home, so some have been tempted by Titan's Earth-like appearance to suppose there might be a refuge there, if and when we exhaust the Earth's capacity to sustain us. Titan has an atmosphere that is mostly nitrogen and with a pressure similar to the Earth's. Beneath its clouds it has a surface of sand dunes and great lakes, a landscape of hills and valleys that is renewed by surface activity. But, though perhaps not quite as bad as Venus, Titan is no paradise. The surface temperature is about −180 °C. There is mist and rain, alas offering no refreshment for earthly life – it is liquid hydrocarbon, cold and oily. We are not suited for life on Venus or on Titan.

Uranus, Neptune, and Outward

Once there were the heavens of fixed stars, with seven wanderers – Sun, Moon, and five others, namely Mercury, Venus, Mars, Jupiter, and Saturn. The Sun took a year to pass through the twelve constellations of the background stars, through the zodiac – Capricorn, Aries, Pisces, and the rest. Occasionally, there were comets with their wispy tails pointing downstream from the Sun, strange visitors giving rise to stranger human beliefs.

But that was a long time ago, when all that people could use to follow the tracks of the planets were their own eyes. Lens-grinding technology made telescopes possible, and they allowed the discovery of Uranus and Neptune. They also made it clear that there was a planet beneath our feet all the time, the Earth, one among eight. Now it is we who whirl through the sky and thereby place the Sun against a changing background.

Telescopes did more. Moons of Jupiter and Saturn, and of Mars appeared. So did asteroids; and eventually a "ninth planet," Pluto. People looked for the "tenth planet," and several candidates came along more or less at the same time: Eris, Quaoar, Sedna, and others.

More comets were seen and some were identified as regular visitors returning predictably. Their pathways led out beyond the orbits of known objects, experiencing the fading light of the Sun on their outward journeys and the fading strength of its gravitational attraction that nevertheless could pull them back in again, a generation or a thousand generations later. They expanded the Solar System out toward the stars.

The current rate of discovery of bodies large and small beyond the orbit of Neptune, and the diversity of these Trans Neptunian Objects, is expanding our consciousness of objects and their revolutions, changing our view of the Solar System and thereby, once again, of ourselves.

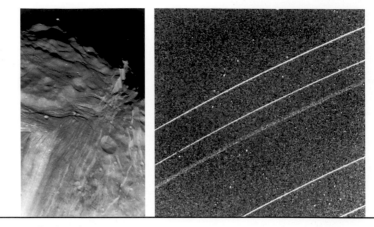

Long dark winter

If you don't like winter, then take care when you visit Uranus. Just as the Earth's tilted axis gives us our seasons, so it is for Uranus. But it has a much bigger tilt so that at northern midwinter the North Pole points almost directly away from the Sun. The deep chill darkness lasts about 40 years.

Uranus has two sets of twins – its largest two moons, Titania and Oberon, are both more than 1,500 kilometers across. Then come Ariel and Umbriel, somewhat smaller but also paired in size. Miranda is smaller, but it was Miranda that the Voyager 2 spacecraft approached most closely almost a generation ago, and so it is Miranda for which high-resolution surface images exist. Voyager gave value for money, and also discovered ten new Uranian moons.

The last planet

Which is bigger, Uranus or Neptune? The former has a greater diameter and the latter has a greater mass. But there is no point in an argument. They are both giant planets.

They are also both blue, somewhat like the Earth, but in their cases it is the gas methane that scatters the light selectively by color, and not nitrogen. Looking for other similarities with our home planet, Neptune may have a core of rock and ice that is Earth-sized. Around that, though, is a huge and windy atmosphere. Clouds can be seen whipping around the upper layers of the planet in just 16 hours, carried by winds that must have speeds up to 2,000 kilometers per hour.

Neptune has one large moon, Triton, which was scanned in detail by the Voyager 2 space-craft. Triton and one other, Nereid, were already known, but Voyager once again found a basketful of new moons, small and dark. It also found Neptune's dark rings, rockier, dustier and perhaps much older than those of Saturn.

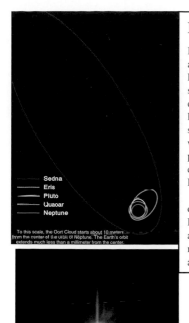

Sedna
Eris
Pluto
Quaoar
Neptune

To this scale, the Oort Cloud starts about 10 meters from the center of the orbit of Neptune. The Earth's orbit extends much less than a millimeter from the center.

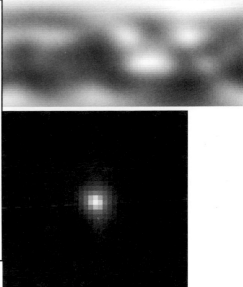

Neither villains nor victims

It is a human habit to impose characteristics and feelings upon inanimate objects. Poor Pluto, we might say, demoted from the grand status of a planet to a mere dwarf. But Pluto doesn't care. And anyway, though it may have lost a title, it has gained moons. Pluto, seen as a single faint dot through Earth-bound telescopes, was always in fact a pair of bodies – the dwarf planet and its large companion, Charon. Then, quite recently, two smaller moons, Nix and Hydra, were found.

Eris is one of the reasons for the downgrading of Pluto. But Eris is no villain to Pluto's victim. In the Solar System, there is ice and rock, there are some gases, and even a little liquid, but mostly there is just terrible emptiness. There are neither villains nor victims.

Poets and scientists

Poets and storytellers who like to play with cloudy metaphors and enticing ambiguities have lost a little something, or so it might seem at first. Telescopic images and close-up spacecraft encounters with comets have taken away some of their mystery.

Now we know where the comets come from, and return to, and what they are made of. We know what they tell us about deep time and deep space, and deep cold. In that very knowledge, though, there is profundity enough for everybody, for poets and for scientists. Mystery has many layers and after each one there is another, perhaps more complex. The purpose of exploration is not to destroy mystery but to find those greater depths. New knowledge does not diminish the joy of imagination, but gives it so much more scope.

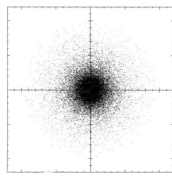

Nemesis and the Oort Cloud

Science works by hypotheses – suggestions based on available information, providing a focus for gathering more data and leading on from there to rejection or cautious acceptance of new ideas. Thus it was when Nemesis was proposed as a dull twin "star" of the Sun. The first data came from subtle periodicity of mass extinctions on Earth – upheavals at 26-million-year intervals. Patterns require suggestions for explanation; they ask for hypotheses. Nemesis, ejecting comets from the remote Oort Cloud to shower inward and create a deadly bombardment, was one such.

Ice, Rock, and Beauty

The Wonder Just Goes On and On

There are some words that should not be used too much. They can lose their impact through overfamiliarity. Wonder is such a word.

But look around, wherever you are. Listen, too. Activate all of the portals to your mind. Open all inward routes through its bony shell. The physical world offers as much wonder as any mind can deal with. Whoever speaks of anything as "only physical" is not sensing very much of what there is in the world to experience. There is nothing "only" about the physical.

We can look at our immediate surroundings, or explore the whole Earth and its interactions, its life. Most of all, we can recognize the wondrous physicality of our own brains and of their actions that give rise to mind and self. The more we extend our raw perception using aids from scanning microscopes to space missions, the more we see that science does not abolish wonder. It magnifies and celebrates it.

Look upward and outward with naked earthbound eyes or with technologies like the Cassini spacecraft. Look at the profile of Enceladus, from within its shadow, and see the sunlight shining through its ice fountains. The wonder just goes on and on.

The Sun

The Earth

Saturn

Enceladus

The object

The ice plumes of Enceladus

The plumes, or fountains, which extend for hundreds of kilometers above the surface of Enceladus (pronounced en-sell-ah-dus) and are feeding material to one of Saturn's faint outer rings, are crystals of ice bursting from what seem to be lakes of liquid water beneath the frozen surface of the moon. Something, as yet unknown, is heating the innards of the moon. More than that, among the ice on the surface fractures and in the plumes themselves, there are simple organic materials. Enceladus is no inert body, but a moon where things are happening.

The image

With the Sun shining from the far side to silhouette the moon but light up its icy fountains, the Cassini spacecraft produced this image from orbit, in 2006.

Credit: NASA/JPL/Space Science Institute

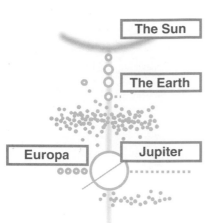

Within Ourselves

The combination of gravity and motion in an orbit that is not a perfect circle produces tensions, and tidal effects, during Europa's 85 hours journey all the way around Jupiter. The strain is showing. Cracks have appeared, seemingly filled with muddier material that has oozed up from deeper levels. The effect, to distant human eyes and minds, is appealing.

Observations say as much about the observer as the observed, and the origins of that appeal, within us, are interesting. For conscious social beings, survival is of diminished value without positive and shared emotions such as love, hope, and beauty. We can, in our heads, attach beauty to Europa's image, and our lives are enhanced.

The object

Europa

One of the four Galilean moons of Jupiter, Europa, has a diameter that is a little less than that of the Earth's Moon. It is covered with ice but probably with a liquid water layer between the ice and the rock below.

The image

This is a false color image, constructed in June 1996 from three separate images made by equipment carried by the Galileo spacecraft, using visible and infrared light. The colors enhance the features for the human eye; the plains of purer water ice are shown in blue, while the red-brown indicates more mixed material. The area of Europa shown is about 1,250 kilometers across.

Credit: NASA/JPL – Caltech

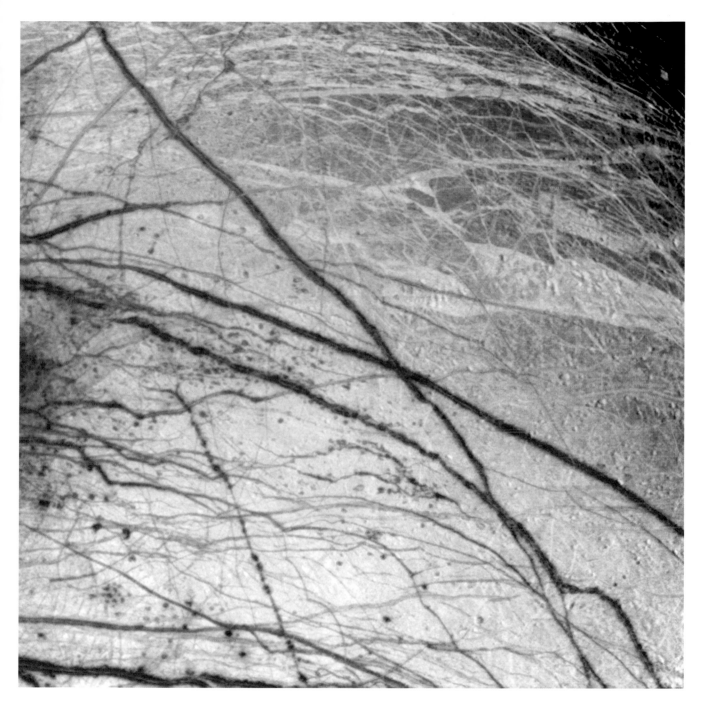

That's You, Right in the Middle

Look at an old class photo from a few (or not so few) years ago, and you usually have to think for a moment about where you are, second row back, third from the right. With this picture it should be easier; you are almost right in the middle. Just a thousandth of a millimeter from the center. The Sun occupies a tiny point at the very center, whereas the most distant point on Neptune's orbit is a small hair's breadth out.

This is not a photograph, of course, but a graph developed from data about the pathways of long-period comets, reaching out to a distance 100,000 times further from the Sun than the Earth's orbit. It is, in other words, a model.

It is a representation of the Oort Cloud. More than 50 years ago it was suggested that those comets which are less frequent visitors to our local skies, the "long period comets," spend most of their time way out, WAY out, beyond the furthest planet; and that there were a lot of them. Observations have continued to accumulate since then, and they all support the idea that there are very many comets out there, perhaps a thousand billion, with a collective mass that could be as big as that of Jupiter. Each dot on the grid here represents such an object, and where they are most numerous the space is solid black. That is a little misleading, since each body is relatively small, and very small indeed compared with the distance from one to the next. But this is a model intended to help scientists to visualize and to create new ideas that can be tested by ongoing observations. It is not meant to be quite the same kind of mirror of reality as a class photo claims to be.

The object

The Oort Cloud

The Oort Cloud is the natural habitat of some of the comets, like comet Hale-Bopp that was clear in our skies in 1997, or the Kreutz sungrazers. These are long-period comets, visible only at intervals of more than 200 years, with orbits suggesting that their furthest distance from the Sun is in the region of 50,000 astronomical units (an astronomical unit being the distance of the Earth from the Sun). They are ancient objects of rock and ice, but with life expectancies that shorten drastically if and when they begin their visits to the inner Solar System where they are blasted by the rays of the Sun. These visitors seem to have been thrown from their more regular orbits, maybe by the gravitational effect of other stars, leaving a swarm of others behind in the cold and dark of the outer Solar System.

The image

The graph, ambitious in its scale since it extends so far from the Sun, was prepared by Piotr Dybczynski using his Oort Cloud simulation software. It provides a summary of the scale of the Oort Cloud and the distribution of comets.

Credit: Piotr Dybczynski

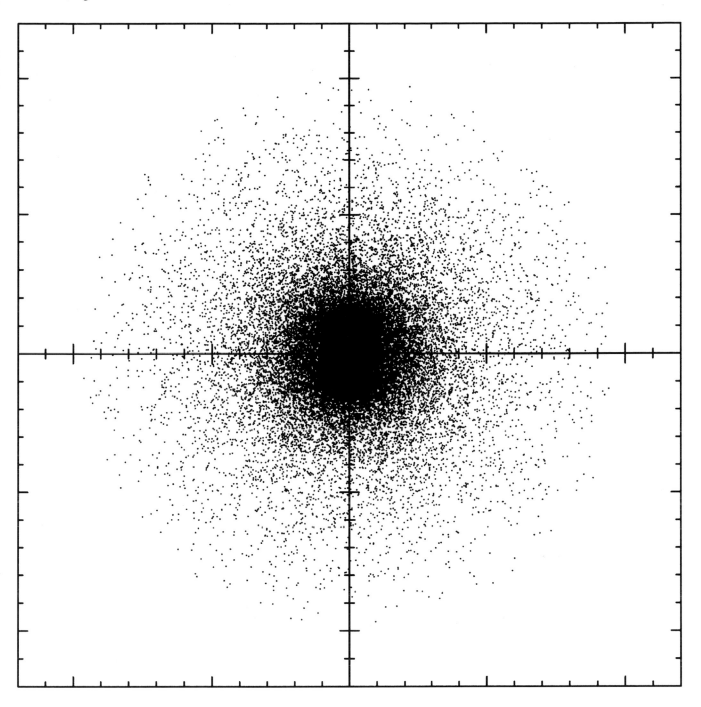

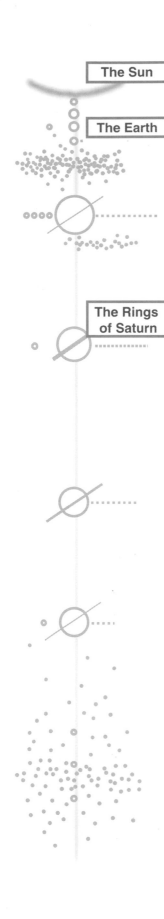

Not Quite a Bee's Eye View

Bees can't see red light, but unlike us they can make their way through the world guided by ultraviolet radiation. We have learned to cheat, however, and to use technologies to do some sensing for us. The ultraviolet imaging spectrograph carried by the Cassini spacecraft, for example, allows us to share a "bee's-eye-view" of the rings of Saturn.

The technologies have here translated different intensities of ultraviolet radiation into different colors. It's another trick. We normally see different colors because of different wavelengths of the light, not because of different intensities or brightnesses. But the use of "false" color makes data friendlier to the human eye. The strongest sources of UV appear in this image as yellow, and the dullest are deep blue. The sky, the Solar System background, is red, thanks to faint ultraviolet emission by low density hydrogen that is found across, and beyond, the Solar System.

It took a long time to produce the image – an exposure time of 9 hours. And in that time, a star, 26 Taurus, crossed the field of view, far beyond Saturn's rings, leaving a yellow trail in the imaging system. Stars, like the Sun, emit UV.

Where the rings are most fragile, as in Saturn's A ring toward the top of the picture, the ultraviolet starlight shines through, while the denser B ring blocks the radiation and its blue arc is almost unbroken.

It seems unlikely, though, that any bee has ever seen rings like this. Bees travel over fields and hills, but not as far as Saturn to get such a close-up view.

The object

Rings of Saturn

The rings together are hundreds of thousands of kilometers across, but only tens or at most hundreds of meters thick. The A and B rings are the most prominent. A is further from Saturn than B, and they are separated by a dark band, a relatively empty strip, called the Cassini gap. The B ring is particularly dynamic – so dense that its grains of ice continuously jostle each other. It is believed that it is the actions of moons, like Mimas, Pan, and Prometheus, that create the gaps and sustain the stability of the ring systems.

The image

The image dates from May 2005, and was made by the Cassini ultraviolet imaging spectrograph.

Credit: NASA/JPL/University of Colorado

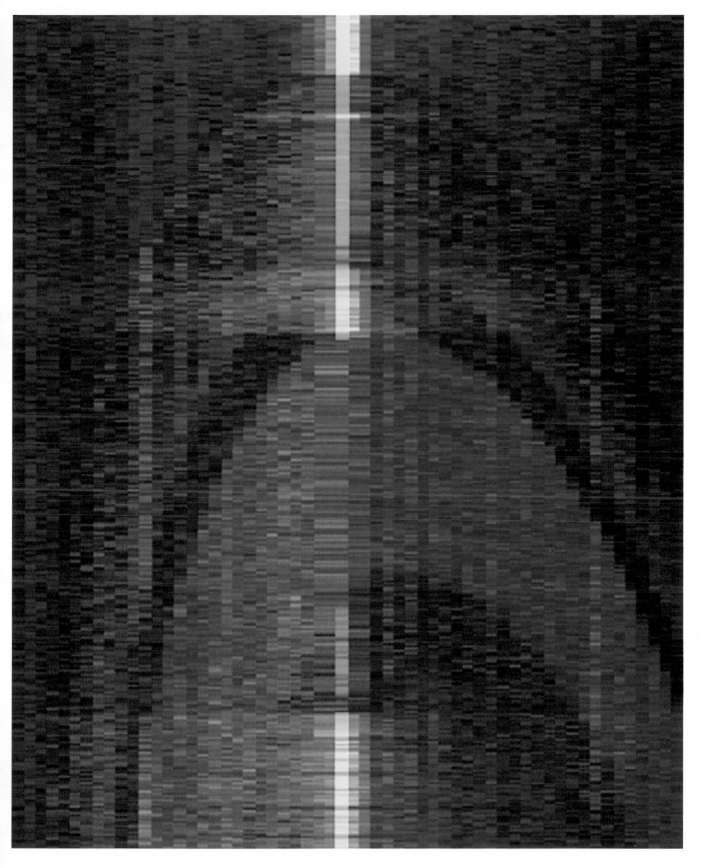

The Namer and the Named

Is it wrong to give names to inanimate bodies, based on human emotion, so that a negative association can become fixed?

Mars, in ancient European mythology, was the planet of war. Careful observers, using eyes alone, can see its bloody color. So, in the nineteenth century, when telescopes had reached sufficient resolving ability, two small satellites were discovered and they were given matching names: Phobos, from the Greek for fear; and Deimos, from the Greek for panic or terror. This image shows Phobos.

Being small and dark, not much was known about these moons until space missions could approach them. It has long been suspected that they are asteroids that have fallen into the gravitational presence of Mars, to travel there for billions of years. Closer inspection made possible by the Mars Global Surveyor mission in 1998 shows that, indeed, Phobos has the appearance of an asteroid. There are craters large and small, some old and dusty, some new and sharp.

But an asteroid trapped in Mars' gravity need not be a cause of fear or panic among Earth-dwellers. There may be asteroids still dangerously at large, capable of catastrophic Earth collision, but Phobos will do us no harm. Its fearful name means nothing. Names, anyway, say more about the namer than the named.

The object

Phobos

Phobos is just 27 kilometers long. Like so many other objects in the Solar System, it is a battered body with craters and cracks. It also has boulders up to 50 meters across, and a thick layer of fine dust that is easily disturbed by new impacts.

The image

Taken on August 19, 1978 from less than a thousand kilometers away, this image shows features that are as small as 4 meters across. At the bottom is the very large Stickney crater, which is so large compared with the small moon that the crater-forming collision must nearly have shattered it. The grooves that spread from the crater rim seem, indeed, to be fractures caused by the mighty impact.

Credit: Viking Project/JPL/NASA/Edwin V Bell II (NSSDC/Raytheon ITSS)

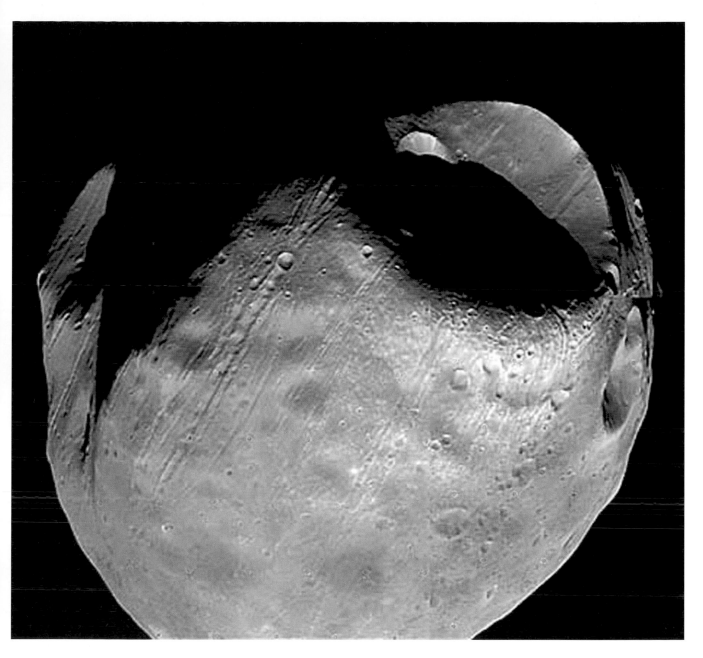

The Truth but not the Whole Truth

The Earth is a ball, but we educate generation after generation to view representations on 2D surfaces. To represent the whole global surface, we sometimes stretch the poles, single points, into lines at the top and bottom of these flat sheets that are as long as the equator. We are so used to such distortions of the truth that we need to be persuaded that the shortest route from New York to Beijing passes very close to the North Pole. We propagate a falsehood that is so absurd that it is a wonder that anyone takes it seriously at all. But we do it because flat surfaces are so very convenient. We can pin them to walls, we can fold them up and carry them around, and we can compile them into books.

The producers of this photographic map of Io have tried to compromise by having the benefit of a flat surface without too much distortion of geometry. They limited the map area to a zone that extends for a little over 60° either side of a central great circle, or equator, rather than something approaching 90° as on most world maps. Even so, there is distortion, and features that do not lie on that central line are extended laterally.

The purpose of a useful representation, however, is to provide some truth, some predictive reliability. They need not always tell the whole truth. Maps do not offer completeness – that exists only in the terrain itself. But they tell us what to expect around the next corner, and their flatness is very convenient indeed.

The object

Io

Of the four Galilean moons of Jupiter, Io (pronounced eye-oh) is closest to the planet, and has the most active surface. Existence so very close to a giant is not free of stress, a consequence of which is that there is little of the surface that is not coated in volcanic dust of some kind or the another.

The image

The Galileo spacecraft provided the raw material for the image, over a period of several weeks. It produced a mosaic of images using infrared, green, and violet light, which have been converted to red, green, and blue versions and then superimposed to produce this final "false color" effect. Active and recent volcanoes are visible as black scars, while the red ring is a result of deposition of volcanic material. The gray, yellow, and orange areas, too, are created by sulfur compounds released from the sphere's interior.

Credit: NASA/JPL/USGS

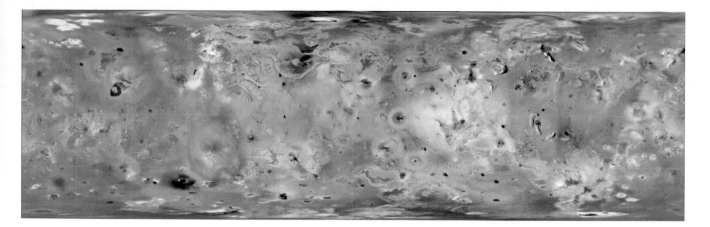

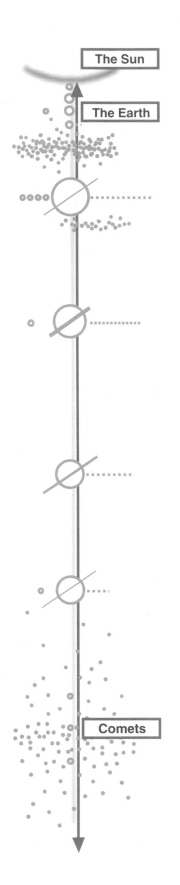

What Happened Next

There are a lot of craters in the Solar System. One of them out there is human-made. Nobody knows quite how big it is – the violent process of making it and then making away from the scene allowed less than 15 minutes for close-up inspection, and in that time the dust thrown up obscured the view. It could be the size of a back yard, or of a football stadium. Whichever, it's now a scar on the face of comet Tempel 1.

It's relatively easy to study the gas and dust coma of a comet, by analysis of the way it reflects and absorbs sunlight and by analysis of the way it sometimes emits light of its own. Finding out what is inside, however, is impossible without disturbing the surface, which is what the copper impactor released from the Deep Impact spacecraft did, hitting the comet's surface at about 10 kilometers per second, and digging through the surface to liberate material from below.

A lot of people watched what happened next, even though it all happened almost as far as the Sun is from us. The Hubble Space Telescope, the Spitzer Space Telescope, the Chandra X-ray Observatory, and ground telescopes all around the world turned toward the impact to help analyze the light and hence identify the materials.

After the Deep Impact collisions, what happened next, as the dust diverted sunlight for inspection by so many watchers here on Earth, tells us more about what happened a very long time ago, about the processes that formed the Solar System, and thus about ourselves.

The object

Comet Tempel 1

Tempel 1 is a comet with a very short orbital period – that is, it takes only five and a half years to go around the Sun. Unlike some comets, it never goes all that far from the Sun (if you count the orbits of the giant outer planets as not all that far). It also still has quite a large nucleus.

The image

The image, taken about a minute after impact on July 4, 2005, shows sunlight scattering off the cloud of dust made by the crash of the Deep Impact impactor.

Credit: NASA/JPL-Caltech/UMD

Up and Down

Asteroid Eros is a billion times smaller than the Earth, in terms of mass, and so its gravitational attraction is accordingly smaller. The pull is enough, though, to have caused fellow travelers in the asteroid belt to swerve if their pathways have come close, and sometimes to crash to the surface, making craters and also the dust that covers the surface. Gravity's influence may be relatively small here, but it reaches out all the same from the asteroid's surface into the surrounding void.

The image shows boulders that seem to have fallen to a crater bottom. Perhaps, though, with the weak gravitational pull of Eros, phrases like "falling to the bottom" don't apply quite as they do on Earth. The experience of a visit to the asteroid would not involve a lot of falling, and there would be only a little sense of up and down. These terms, "up" and "down," are useful to us in our local habitat; we live oriented with a stronger force of gravity. No such orientation is possible at all in free space, and on Eros it is a minor effect; any sense of verticality there would largely be a result of the object's visual presence.

The object

Eros

Eros is an "Amor" asteroid, one that crosses the orbit of Mars and reaches inward almost to the Earth's pathway. In 1931, it passed within 23 million kilometers of our planet. It is now a much-studied body, since the NEAR-Shoemaker spacecraft spent a year in orbit through 2000 and into 2001, when it made a final landing on its surface.

The image

The picture was taken from just 50 kilometers away by the NEAR-Shoemaker robotic spacecraft, and shows the largest crater on Eros.

Credit: NASA/JPL – Caltech

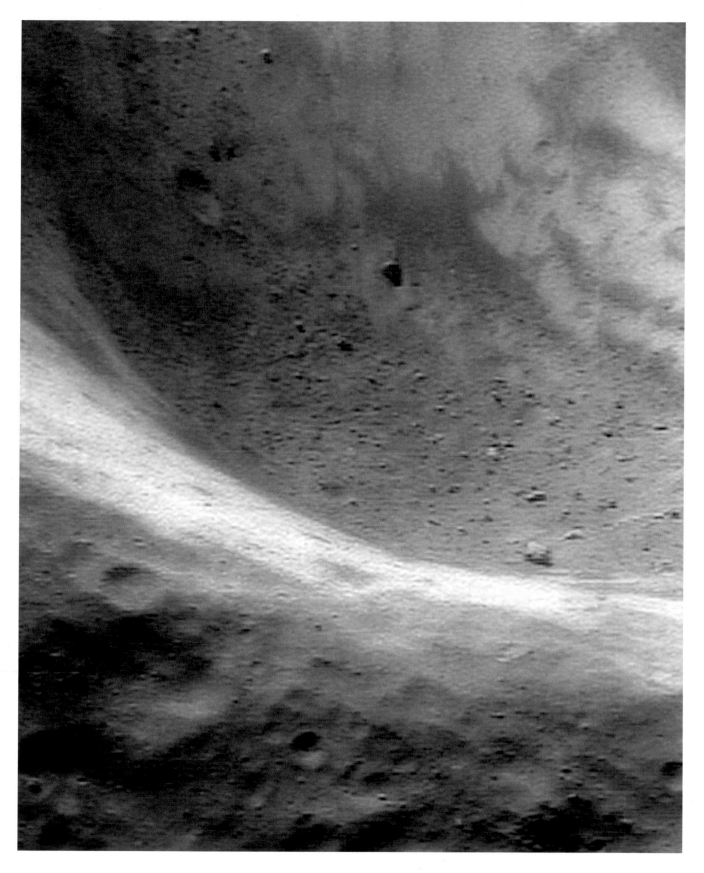

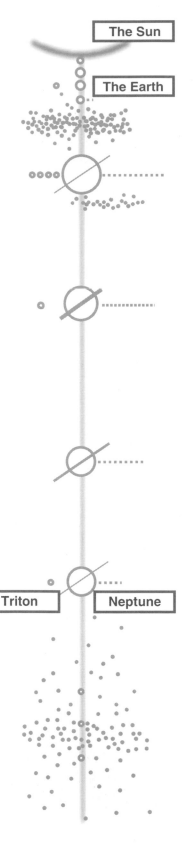

What It's Good At

Science predicts. A scientific idea lives or dies by doing so. Predictions that turn out to be wrong suggest false principles – and only the best ideas survive. The result is knowledge that is reliable, but not absolutely infallible. It stands ready to be falsified by new observations, by newly failed predictions. It is knowledge, not dogma.

Science can predict the existence of entire planets and moons. So it was with Neptune. Following the discovery of Uranus more than 200 years ago, astronomers were keen to predict the planet's motion for years to come. But their expectations didn't quite work. First it went just a bit too fast, and then slightly too slowly. One possible reason was that the foundations of prediction were wrong; that the laws of gravity were not true laws. Another was that something unobserved was pushing and pulling, just a little, on Uranus.

Astronomers worked on both possibilities. Some developed an interesting idea that there was an invisible planet out beyond Uranus, tugging on it with its gravity, speeding it up just a little or slowing it down. They were even able to say roughly where this "imagined" planet would be, and persuaded others to look in that direction. And there it was – the fourth largest planet in the Solar System, a gassy giant – Neptune. Just a few weeks later Triton was spotted, Neptune's largest companion, one of the larger moons of the Solar System.

Science predicts. That's what it's good at.

The object

Triton

With a diameter of 2,706 kilometers, Triton is the only one of the large moons of the Solar System that orbits in retrograde fashion – in the opposite direction to the spin of the planet. This suggests that it is an object captured by the already-formed planet, rather than one that shared its origins with Neptune. The nature of its orbit may also be linked to tidal effects that stress and heat the moon and explain the observation that the surface is still active. Triton has a cycle of seasons that lasts hundreds of years, producing slow changes in the polar caps of nitrogen ice and in the moon's thin atmosphere.

The image

After the Voyager 2 spacecraft had made its closest approach to Neptune in 1989 it looked back, from nearly 5 million kilometers away, to obtain this image of Triton dwarfed by Neptune.

Credit: NASA/JPL – Caltech

Third Rock and a Bit

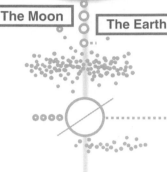

Working outward from the Sun, the first body that is a close companion to another, traveling in partnership around the central star, is the Earth's own Moon.

Note, in passing, the astronomers' reverence. It is not merely the third rock from the central star, nor just the earth, but it is the Earth. It is not the moon but the Moon. There are very many moons in the Solar System, but only one Moon. Likewise it is the Sun and not a mere sun that shines on the planets, Mercury, Venus and the rest. Astronomy is an ancient practice, and though observations have often required old ideas to be abandoned, it has lost none of its ancient respect.

Our own satellite is not the largest in the Solar System, but there are only four that exceed it, and these orbit much bigger planets than the Earth. The Earth and Moon are a dancing couple, not quite equal in size but matched well enough for an ancient and endless pas de deux under the light of the Sun. Respect is due to them, just as it is to every planet and every moon.

The object

The Moon

The Moon's diameter is 3,476 kilometers (while that of the Earth is 12,750 kilometers). Its mean surface temperature is very variable, between −175 °C in darkness and +125 °C on the sunlit side. It has a density that is larger than that of the Earth's crust of solid rock, but significantly less than that of the Earth as a whole, suggesting a core that is less dense than the Earth's. (This is in agreement with the observation that the Moon has some local magnetism but no overall magnetic field and probably lacks the same iron-rich center as the Earth.)

The image

Dating from September 22, 1992, while on its journey to Jupiter, the Galileo spacecraft took images with three filters, which here are superimposed. And though the light was the usual reflected sunlight, the combination can be shown with added false color to emphasize differences in mineral types of the Moon's surface.

Credit: NASA/JPL – Caltech

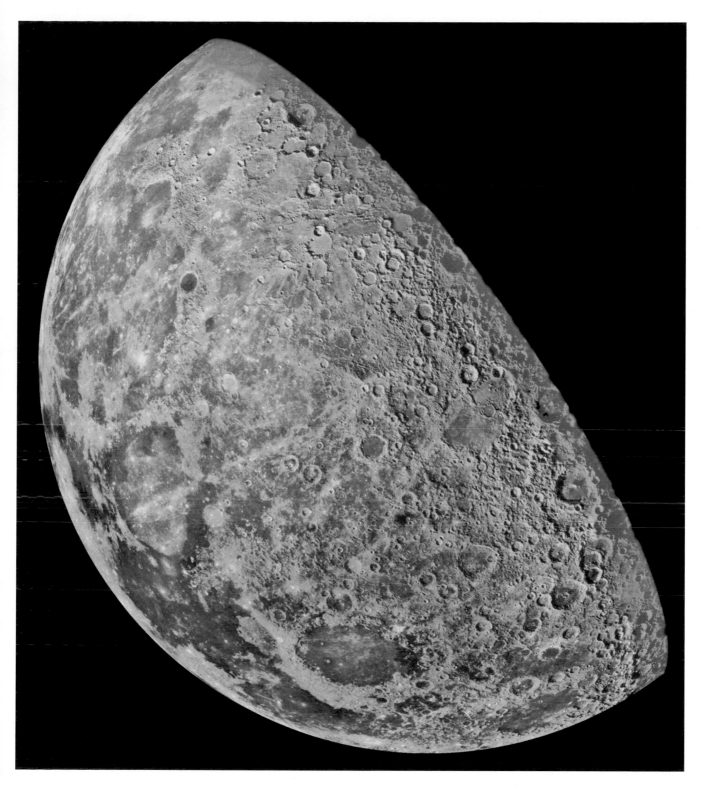

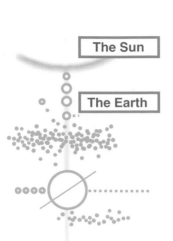

The Sun

The Earth

Saturn

Enceladus

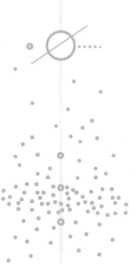

A Fabulous Night

The regular transitions of the Moon from crescent to globe and back again, as its sunlit half shifts in and out of our view, impact less on electric city existence than on the lives of our ancestors. Many of us have left the Moon behind and moved on. The night, however, remains central to ordinary life.

Night is a local affair, a matter of spin. As passengers on one of the Sun's planets, we too must spin so that the rhythms of our lives run in time with the rhythms of the Earth's particular dance.

Out among Saturn's many satellites the Sun has less intensity than in the cosier band of the Solar System where we live, and a bright midday is never going to be so very bright on a body such as Enceladus. But as on Earth, there is spin and thus there are cycles of midday, twilight, and midnight. As it happens, the time for this moon of Saturn to make one rotation is not so different to the pattern that we experience here at home. On Enceladus, we wouldn't like the cold (or the vacuum or the radiation). We'd be surprised by the sunlit bands across the night sky, the rings of Saturn. But at least our body rhythms would be in tune with the local daylight.

The object

Enceladus

Enceladus has synchronous rotation, meaning that the time for one spin around itself is the same as the time for one journey around Saturn, which is not that different to the time for the Earth's daily turn. Enceladus, however, is much smaller than the Earth, and has no atmosphere to create the kind of effects we experience here at home – blue skies, clouds, rainbows, red sunsets. But it has its crystal fountains. And being made of fairly clean ice, it offers other visual excitements. It returns the sunlight that hits it right back out into space, so that it is a bright and shiny place.

The image

Enceladus is seen in front of the thin rings of Saturn across which the shadow of the planet is falling to create the tapering effect. It is a Cassini image, taken from a little over a million kilometers from Enceladus, on March 7, 2005.

Credit: NASA/JPL/Space Science Institute

Aorounga to Zapadnaya (by Way of Manicouagan)

There is a planet that scatters blue light from its star back out into space. It is a world with only a few craters made by impacts of smaller bodies. Its atmosphere offers protection to its surface, at least from the smaller specks of material that fall inward from space. Much of its surface is covered with water, so that an impact usually makes no more than a splash. And its rocky zones are in motion, ever-changing, so that no surface feature, however momentous the event that creates it, lasts for very long. But there are craters, just a few that are clearly recognizable as such and all of them very much younger than the planet itself.

The planet has, among its complexity of life, some inhabitants who have given names to the craters they have so far discovered. These names show local variations. There is the Aorounga crater in a desert region, and Zapadnaya in the heart of a fertile continent. Elsewhere there are El'gygytgyn and Tabun-Khara-Obo, Goat Paddock and Tookoonooka, Rio Cuarto and Manicouagan. It is a planet rich in names.

The objects

The Earth

The planet is Earth, the largest of the Solar System's rocky spheres. It is largely but by no means entirely covered with water, with an atmosphere that has an exceptionally high oxygen content due to the energy capture processes of most of its living material.

Manicouagan crater

The feature in the image is Manicouagan crater in Quebec, Canada, approaching 100 kilometers in diameter and easily visible from space, aided by the presence of a ring of water. As with other craters, the rocks around it include material that did not originate in the ground below, and so must have come from above. Further study of the rocks indicates that the impact took place between 213 and 215 million years ago. The circular lake is in fact a product of the activities of planetary inhabitants (resulting from a hydroelectric installation).

The image

The image, from the spring of 1992, shows the Manicouagan crater and the "limb" of the Earth where the blue atmosphere fades into the black vacuum of space.

Credit: Nasa Johnson Space Center - Earth Sciences and Image Analysis (NASA – JSC - ES&IA)

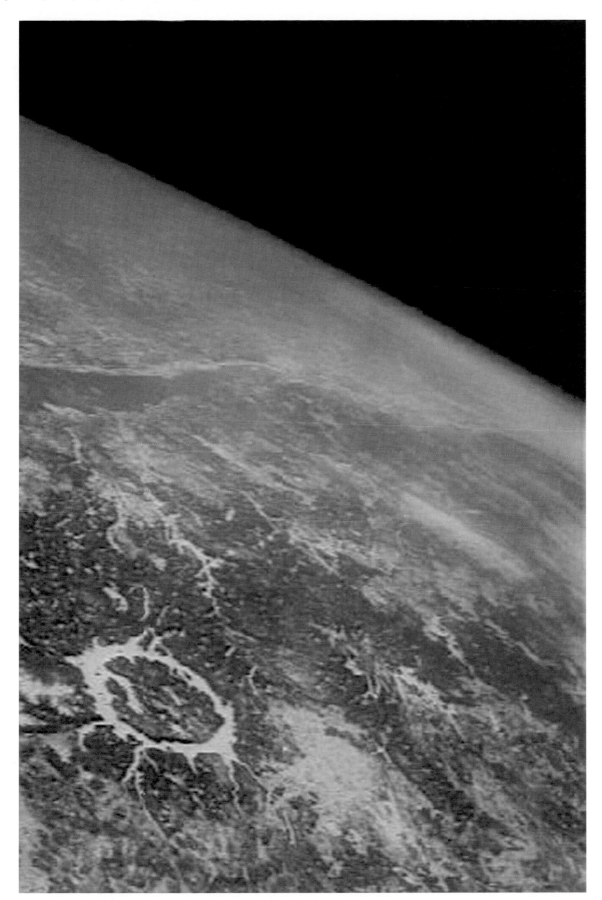

The Sun

The Earth

Saturn

Hyperion

Known and Unknown

It might be possible to describe Saturn's moon Hyperion as an oddball, if it were a ball at all. It is the largest object in the Solar System that is not, at least approximately, a sphere. For anything else of the same size, gravity has pulled and crushed material so that surface irregularities are small compared with the whole body. But Hyperion is one big irregularity, in every way.

Its low density suggests that its appearance of a sponge or a cinder is carried down into its interior. It is a moon full of holes.

Its orbit and spin behaviors are every bit as eccentric as its shape and its structure. The time for one journey around Saturn varies from one orbit to the next, thanks to the influence of its larger neighbor, Titan; and its axis of spin tumbles chaotically.

Thus, strange and distant Hyperion highlights the diversity of the many objects that travel around the Sun, either singly like Mercury and Venus, or together in countless hordes, like the asteroids or like Saturn's moons and rings. We are learning rapidly about our solar neighborhood and also about the existence of disks and planets that orbit other stars. Thus Hyperion, as a unique representative of all the known objects, gives rise to thought about how much more diversity lies unknown in the greater beyond.

The object

Hyperion

Hyperion has dimensions of 370 × 280 × 225 kilometers, and a density of 569 kilograms per cubic meter – similar to the density of some timber. A wooden moon would be an interesting first, setting some serious challenges of explanation, but it is not the case for Hyperion.

The image

The Cassini spacecraft flew past Hyperion on September 26, 2005, and its narrow-angle camera produced this image from a distance of 62,000 kilometers. The width of each pixel corresponds to a distance on the moon of about 360 meters.

Credit: NASA/JPL/Space Science Institute

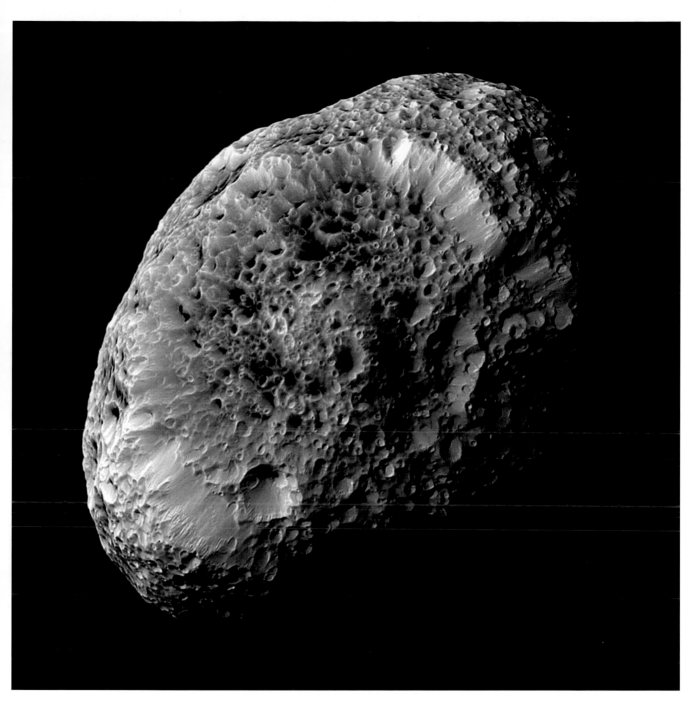

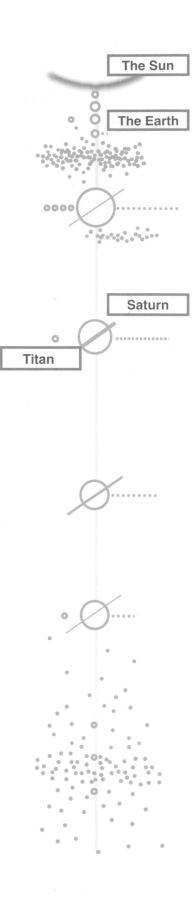

The Sun

The Earth

Saturn

Titan

Works of Nature...

This could, almost, be an imaginative representation of our own Moon. Human imagination and ingenuity are certainly wondrous, but they are matched by works of nature further from home. Titan, largest moon of Saturn, is such a product, and now, thanks to the Cassini-Huygens mission, we can take a close look.

Titan is a distant world discovered centuries ago despite being dwarfed by its host planet. Hidden in thick haze, it may also seem largely unknowable. But the Cassini spacecraft has mapping systems that can penetrate the atmosphere, and the Huygens lander, which accompanied Cassini on its seven-year journey from Earth, parachuted to the surface. There, we must assume, it now lies passively on the surface, a little fragment of human ingenuity a long way from home. By such means we have begun to see through the depths of Titan's mists.

Nature's depths, relative to our knowledge of them, are bottomless. We exist, like Titan and everything else in the Solar System, among those depths. One of the joys of living is to be part of them, and so to be able to experience them, however incompletely. We are not outsiders looking in, but nature gazing as best we can upon nature.

The object

Titan

Titan is rather smaller than our own Moon. It has a density of about 1,900 kilograms per cubic meter, less than that of familiar rock, so it is probably a mixture of rock and water ice. The atmosphere that produces the interesting haze is mostly nitrogen, just as for the Earth, but with hydrocarbons such as methane and compounds of hydrogen, carbon, and nitrogen. Unlike here on Earth, there is no oxygen.

The image

The Earth's sky is blue, while Titan's orange-brown atmosphere is topped by a fragile haze of purple. In this image that layer is made more clearly visible by superimposing an image made using ultraviolet light upon one made using filters that select visible light – red, green, and blue. The data came from the Cassini spacecraft, on May 5, 2005, when Cassini was 1.4 million kilometers from Titan.

Credit: NASA/JPL/Space Science Institute

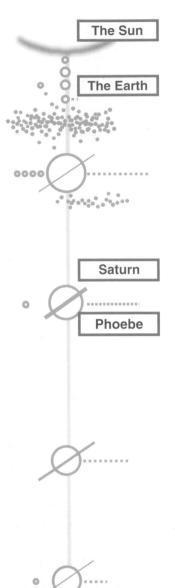

The Sun

The Earth

Saturn

Phoebe

...And Cold Indifference

Is Phoebe a natural child of Saturn, or a foundling?

Phoebe is a small moon, just over 200 kilometers across, made of more stone and less ice than Saturn's inner moons, adding credence to the suggestion that it is not quite part of the family, but something of an outsider. One possibility is that it is a captured Centaur, an ice-rock that once wandered more freely through the realms of the gas giants until falling into Saturn's gravitational grasp. If so, then it is the only such object that has been imaged in anything like the detail seen here.

Whatever its origins, it has known hard times. The surface is pitted with the scars of bombardment. Phoebe, it seems, is a lonely battered child. We can experience beauty, but we are wise if we do so in full awareness of the scale of nature's cold indifference to the fates of moons and of children, women and men.

The object

Phoebe

Phoebe is a long distance from Saturn, 12.9 million kilometers, and takes about a year and a half to make one orbit, though it spins around its own axis every 9 hours. Its craters show bands of light and dark, rather as if impacts have stirred the dark upper layers of the moon and exposed the cleaner material below.

The image

This is a Cassini image from 2004, when the spacecraft was on its approach to Saturn and had not yet gone into orbit. Cassini was able to make a close fly-by, passing little more than 2,000 meters from Phoebe, and thus was able to produce such a detailed image of the surface.

Credit: NASA/JPL/Space Science Institute

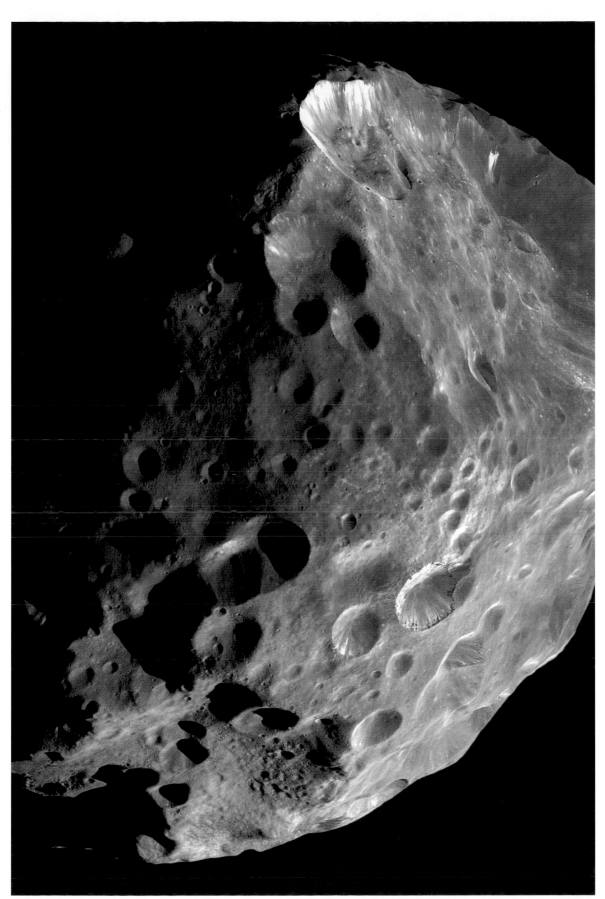

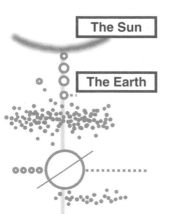

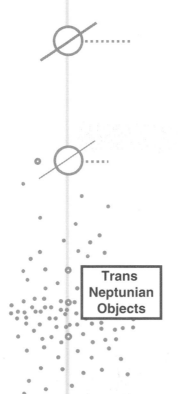

Enough Way Beyonds

There are a lot of "way beyonds" in astronomy. Most of the asteroids are way beyond Mars; Neptune is way beyond Jupiter, Saturn, and Uranus. After that come the Trans Neptunian Objects.

Trans Neptunian space is anything but empty. It contains the Kuiper Belt of objects that include Pluto and Charon and their smaller moons. There are Varuna (after a Hindi god), Orcus (a Roman god of the dead), Ixion (from Greek stories, a murderer chained forever to a burning wheel), and Quaoar (a much nicer sort who, according to one Native American creation tale, danced with the Sun and the Moon to bring the world into existence from its previous chaos). There are plenty more, named and un-named.

Scattered beyond the Kuiper Belt is a disk of objects such as Eris and Sedna. The latter is certainly way beyond Pluto, and appropriately has the name of an Inuit goddess from the dark and cold depths of the Arctic Ocean.

The Oort Cloud is different again, not a belt or a loop or a disk, but a sphere of about 1,000,000,000,000 bodies surrounding us. Way beyond the Oort Cloud are the nearest stars, and way beyond them are more stars, then some more. But this book is about our own little corner, the Solar System, and that is enough way beyonds for now.

The objects

Pluto, Quaoar, Eris, and Sedna

Pluto was once the "ninth planet" and the search was on for planet number ten. Newspapers reported Quaoar as this tenth planet, and did the same for Sedna. The position was complicated by still more discoveries, notably that of Eris in 2005, larger than Pluto. After heated debate, Pluto was demoted to the status of dwarf planet. Pluto, Quaoar, Eris, and Sedna are representatives of a very large number of bodies, the Trans Neptunian Objects (or TNOs). Approximate diameters of Pluto, Quaoar, Eris, and Sedna, all TNOs, are 2,300; 1,200; 2,400; and 1,600 kilometers respectively.

The image

The image gives an impression of the elliptical nature and relative sizes of the orbits of named Trans Neptunian Objects. The Kuiper Belt is a disk of bodies, and its position corresponds very roughly with the orbit of Pluto. Further than that is a scattered disk of bodies. Sedna has been referred to as an object of the "inner Oort Cloud," but the main cloud that forms a spherical shell around the Solar System seems to start much further away.

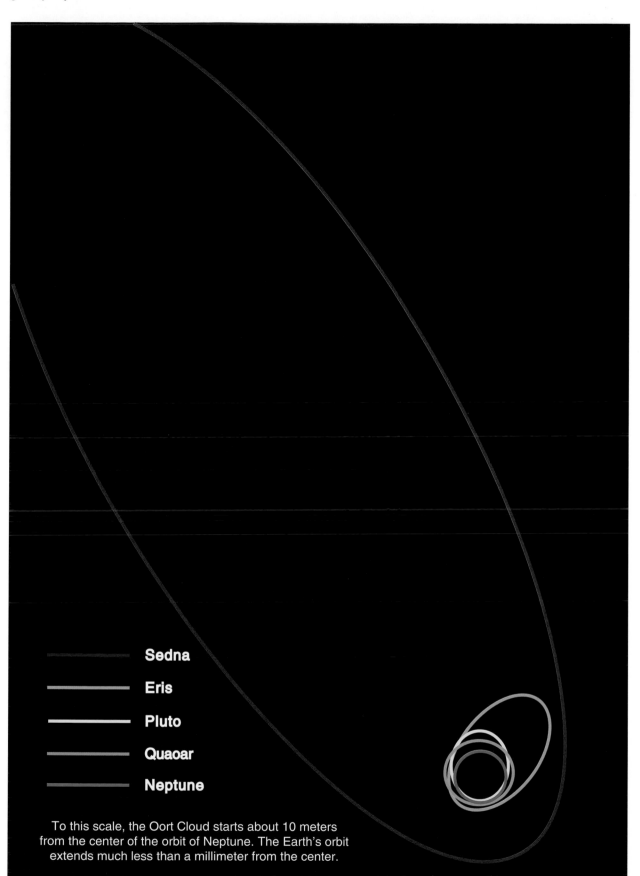

Sedna

Eris

Pluto

Quaoar

Neptune

To this scale, the Oort Cloud starts about 10 meters from the center of the orbit of Neptune. The Earth's orbit extends much less than a millimeter from the center.

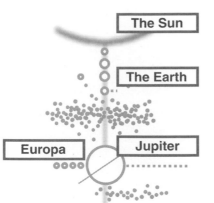

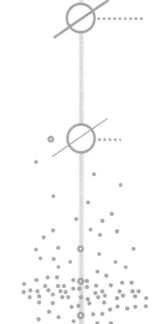

A Pretty Face but a Cold One

Earth and Europa have something in common – a layer of water over and under much of their surfaces, whether in liquid or solid form.

On Earth's surface we have ocean waves and flowing rivers, we have rain and life-giving moisture among grains of soil, and in more solid form there are snowflakes and icicles, beautiful hoar frosts, glaciers, and the ice caps of the poles. Europa's liquid lies below the surface, while on its upper layer the ice makes patterns on a big scale. Its shell of frozen water could be 100 kilometers thick.

A frozen object usually means one that is static, trapped in time. Europa's ice, though, is better compared with the Earth's rock than water. The Earth's crust is rock-on-the-move above a softer layer; dynamic rock subject to judders, cracks, and deeper flows. Europa's is dynamic ice.

It cracks, it melts under pressure, and it heaves upward and falls back. It partially melts and mixes with rock, creating patterns that are anything but bland. Europa has a pretty face, if a cold one.

The object

Europa

Of Jupiter's four largest moons, the Galilean moons, Europa is the second closest to the planet. It is close enough for the gravity of Jupiter to combine with its own non-circular motion to create stretching and squashing effects – tides in the ice – and though the tidal effect is not as strong as on sister-moon Io, these help to produce the cracks and the general dynamism of the surface despite the thickness of the frozen layer.

The image

The image shows an area of the moon that is more than 300 kilometers across – the size of a small country. It was taken by NASA's Galileo spacecraft on June 27, 1996.

Credit: NASA/JPL – Caltech

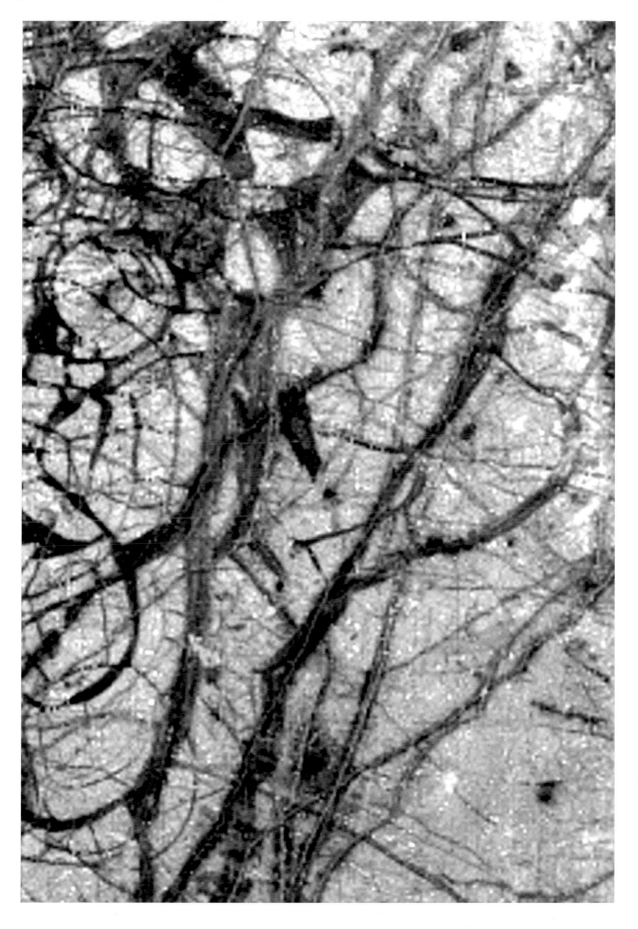

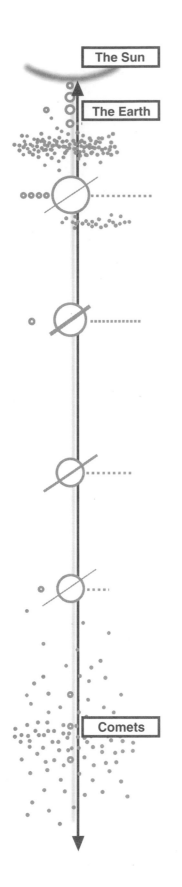

A Journey of a Thousand Lifetimes

It's a long, long story; 75,000 years long. It's the tale of the orbit of comet Kohoutek around the Sun, a journey of a thousand lifetimes.

Its visit to us here in the inner Solar System in the mid-1970s led members of one cult in the USA, The Children of God, to flee to communes around the world, convinced by their leader, David Berg, that doomsday was coming to America. But that's another story – a short one, but intriguing nonetheless in its revelations about the many-faceted relationship between the Cosmos and its people.

Following its first sighting, observers hoped for a spectacular display, from what they thought could be a comet of far distant origins making its very first appearance so close to the Sun, and experiencing its first blasting by the solar wind. In the event, however, the comet was quite faint. It seems that Kohoutek was more rocky than it might have been, less icy, and not subject enough to the assault of the Sun's beams for growth of an impressive halo of loose matter around it.

Those beams did, however, drag enough material away from the comet to create its tail, always ribboning away from the Sun. Now the comet has returned to the darkness, providing a metaphor for others who leave us behind.

The object

Comet Kohoutek

The comet, named after the Czech astronomer who discovered it as it approached us in early 1973, was visible with the naked eye, showing both its nebulous shape around the core and its long tail of material streaming away from the Sun.

The image

Unlike most of the images in this book, this photograph was taken using a traditional 35-millimeter film camera, on January 11, 1974, at the Catalina Observatory of the University of Arizona.

Credit: NASA Johnson Space Center (NASA-JSC)

A Moon of One's Own

The Sun

The Earth

Ida and
Dactyl

Asteroid Ida is a little more than 50 kilometers long and 20 kilometers wide – bigger than a mountain, smaller than a mountain range. Its gravity is not strong enough to pull irregularities into itself as happens for the Earth, where the attraction of matter to matter will not tolerate much departure from the spherical. Here, mountains sink under their own weight and the peaks of the Rockies and the Himalayas cannot be higher than they are, but the shape of the little asteroid is not so constrained.

Ida has many fellow travelers, some bigger and very many smaller, in the Main Belt of asteroids out beyond the pathway of Mars. But it has one very special companion – its own moon, Dactyl.

It isn't necessary to be all that big to start attracting followers. Ida and its satellite experience enough of that strange mutual pull for which possession of mass is the qualifying requirement, so that neither of them is doomed to travel in endless solitude.

The objects

Ida

Ida's density is similar to that of the Earth's rocky crust, while its color suggests a high proportion of metal-rich silicates. Its inner structure seems to be a conglomeration of discrete parts crushed together by gravity, rather than a continuous single body.

Dactyl

In Greek mythology, the Dactyls were small demons, not dissimilar to dwarfs, who mined Mount Ida and were known for their metalworking skills. Dactyl is 1.5 kilometers across – a dwarf indeed compared with Ida but still a substantial piece of stone, by human standards.

The image

The robot spacecraft Galileo generated this image in 1993 as it traveled out, Jupiter-bound, through the Solar System. In doing so, it made the first human discovery of a moon of an asteroid. Since then many more moons of asteroids have been found.

Credit: NASA/JPL – Caltech

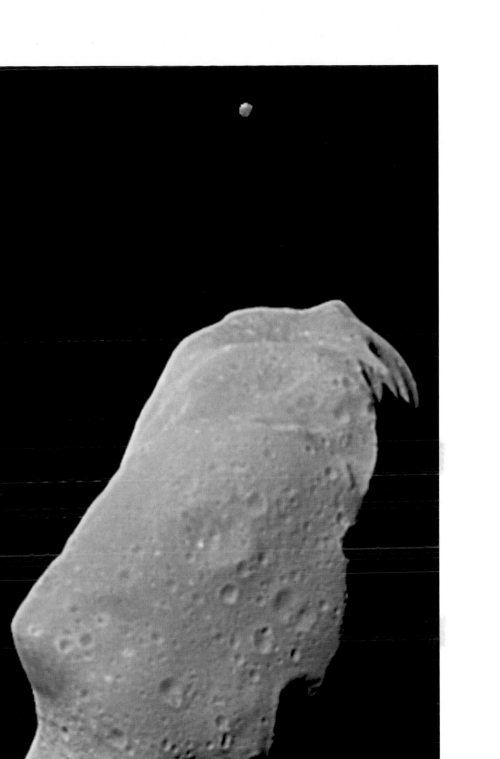

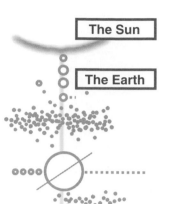

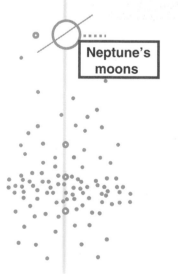

No Substitute for Going There

The Hubble Space Telescope is known for its high-resolution images of distant objects that are far outside the Sun's small family of planets. Those distant galaxies are not just a long way away; they are also very large. Hubble can produce images of our own neighborhood, as here, in which the objects are much smaller as well as much nearer. Neptune has a diameter that is much less than a millionth of the distance across the whole Solar System, which itself is tiny compared to the distance to the nearest star, and tiny again compared with a galaxy.

The image is a useful one, showing some detail on the surface of the planet and also the four moons Proteus (the brightest, at the top), Larissa, Despina, and Galatea. But compare it with the close-ups taken by the Cassini spacecraft in its orbits around Saturn and its fly-bys of its moons, or indeed with the image of Neptune's own moon, Triton, as pictured by Voyager 2 during its fly-by back in 1989. Clearly, in planetary astronomy, there is no substitute for going there.

The objects

Neptune's moons

Experience with Saturn suggests that it is hard to count the number of moons a planet has, since small moons can be discovered at a remarkable rate. A recent count for Neptune is of 13 detected moons. Proteus, Larissa, Despina, and Galatea are all too small to be spherical, but there is also one large and round moon, Triton, which does not appear here.

The image

The image was obtained by the Hubble Space Telescope in April 2005.

Credit: NASA, ESA, E Karkoschka (University of Arizona), H B Hammel (Space Science Institute, Boulder, Colorado)

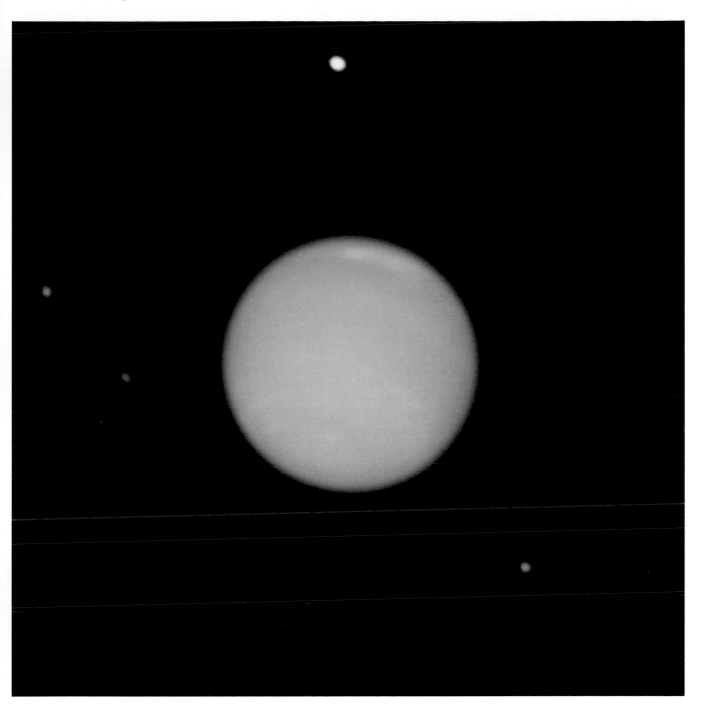

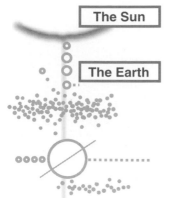

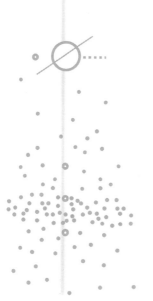

Ringshine

An arc of Saturn shines brightly in the sunlight, against the blackness of space beyond it to the right, and its own night time on the left.

Night time on Saturn is not always quite like the Earth's hours of darkness. On Earth we sometimes have the Moon, as well as the light with which we ourselves flood our planet to prevent our dark side from being truly dark. Saturn has her rings.

Just as the Moon does for the Earth, the rings reflect sunlight to the planet, so that in regions facing the sunlit side of the rings, lower left in this image, the night is softened. It is bathed in ringshine, and if you could look up from the surface you would see bright bands across the sky.

On Saturn, as on Earth, there is twilight, a process of passage between sunward and outward. And there, suspended above the transition as seen from the Cassini spacecraft high above, is a special feature of this special image. It is a gray dot – a small moon, little Epimetheus moving rightward across the image, soon to be wiped from view, swamped by the brightness of Saturn behind it.

The object

Epimetheus

With a maximum diameter of just 115 kilometers, Epimetheus is smaller than some of the craters on some of its bigger-sister moons. It is an irregular object, lacking the gravitational clout to pull itself into a spherical shape.

The image

This is a visible light image made by the Cassini narrow angle camera on June 9, 2006.

Credit: NASA/JPL/Space Science Institute

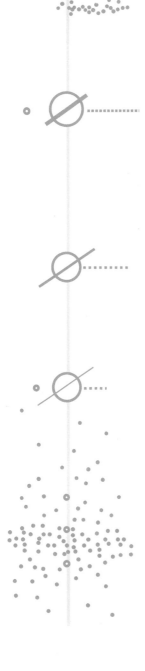

Beneath the Blowing Sands

Mars is a dead planet, a desert ball. It is not quite certain that it has always been so, but if there is life there now it is being very coy – various missions including the Mars Rovers have found no sign. Perhaps one day humans will land there, with sealed spacesuits standing in for the Earth's protecting atmosphere. For now, Mars spins and tumbles around the Sun, carrying nothing but its red dust blown by thin winds that nobody will ever feel on their skin. The planet is closer than the Earth to the main asteroid belt, and there are plenty of asteroids that cross its pathway. Collisions are inevitable. So beneath the blowing sands of its deserts there are craters.

The image here shows a crater eroded by whipping sand, half buried in red dust. But it is not on Mars. It is in the Sahara Desert.

The outer circular depression is more than 12 kilometers across. A large city could fit inside it. Whatever made such a mark must have been big. Not only that, it was probably not alone. There are two other craters, not so easy to see, nearby. This may have been a multiple collision, rather like the collision of the Shoemaker-Levy 9 comet with Jupiter in 1994. Mars is not the only planet that has interacted with the smaller bodies of its neighborhood.

The object

The Aorounga crater

More than 300 million years old, this crater is one of very many on the Earth, disguised by our planet's active geology. We can only suppose that many more have hit the oceans and the ice caps, leaving little lasting trace. An object, whether asteroid or comet, of the size of the Aorounga missile, hits the Earth every million years or so.

The image

The image was taken from Earth orbit in December 2002.

Credit: NASA Johnson Space Center – Earth Sciences and Image Analysis (NASA-JSC-ES&IA)

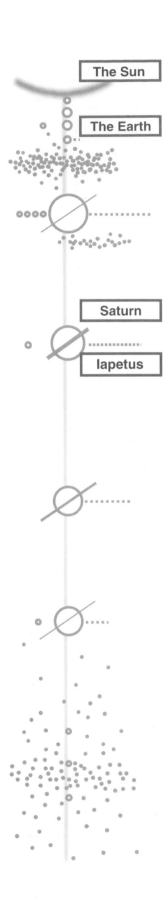

Designed by Committee

If a worthy committee were to design a moon, then some decisions would be relatively straightforward. Yes, it should be round, since that is, after all, the default shape for moons, at least the larger ones. And there should be craters. Moons are supposed to have craters.

But from there on, things become more complex. How big should the craters be? After considerable discussion, the safest option seems to be a mixture of large and small, but just to be absolutely confident it might be best to have at least one crater that is very large indeed, never mind that it then covers a large proportion of the body's surface. How well defined should the craters be – sharp so that they appear young, or eroded by more recent bombardments as would be the case for older features? Again, settling for a mixture avoids too much debate.

Surface brightness, its ability to reflect the light of the Sun, is clearly a contentious point. To please everybody, one half of the moon will have to be shiny bright, and the other deeply dark.

Then there is the difficult matter of the equator. Every moon should have one, of course, a band going all around. A forceful voice stresses the importance of equators in general, and how they are underemphasized in existing moons. A large ridge all around the moon would put all of that right. By now the members of the committee are weary, and agree, no matter that this gives the moon an odd appearance.

As for a name, well, a subcommittee will have to search the catalogs of ancient Greeks and Romans, as usual.

The result could be something like Iapetus – a rather strange place.

The object

Iapetus

The leading hemisphere of Iapetus is extremely dark, reflecting only about 4% of the sunlight that falls onto it. It seems to be a coating, rather than anything with deep structure, and there are various possible explanations. It could be material from within the moon, erupted from volcanoes. Or it could be material that has been blasted from the surface of neighboring moon Phoebe, or another Saturnian companion, by old collisions. Whichever, the surface here contrasts strongly with the other half of the moon, which reflects more than 60% of the sunlight. Only further observation, further evidence, can establish a reliable theory. The equatorial ridge, too, is unexplained, pending further exploration.

The image

The image shows, mostly, the darker leading edge of Iapetus, with the equatorial ridge just visible at upper right. It was taken on November 12, 2005 by the Cassini spacecraft.

Credit: NASA/JPL/Space Science Institute

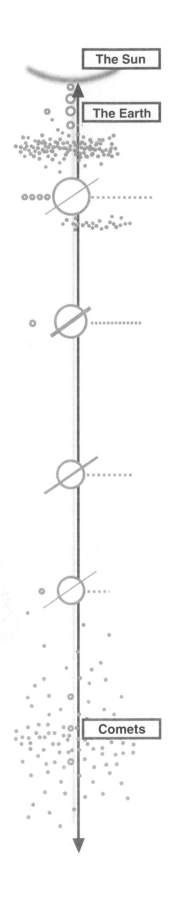

The Sun

The Earth

Comets

No Ordinary String

Comets come closer to the Sun to die. The Sun blasts away at their icy cores; it diminishes them little by little, and sometimes, in the end, it pulls them in and consumes them.

Comet Shoemaker-Levy 9, on its final journey, didn't get as far as the Sun. As it traveled inward through the Solar System it became ensnared in the gravity of giant Jupiter, and broke up into a string of fragments. It was no ordinary string, being more than a million kilometers long, made up of objects as big as mountains. And there was nothing ordinary about what it did.

Over six days in 1994 the fragments smashed one by one into the gassy depths of Jupiter, creating fireballs and a band of dark scars as the planet turned. The biggest of these was as large as the Earth, and persisted for almost a year despite the swirling of Jupiter's clouded layers. For the first time, people witnessed and recorded a collision of a smaller body with a larger one – just the kind of event for which there is so much evidence across the Solar System, on the faces of the more solid planets and moons.

The object

Comet Shoemaker-Levy 9 fragments

The comet was discovered in March 1993, and its pathway soon showed that it was in Jupiter's orbit, having passed very close to the planet surface the previous year, which is probably when its break-up occurred. It may have been in Jupiter's orbit for several decades, but its origin beyond that is unknown. Multiple collisions with Jupiter created a band of slowly fading scars around the fast-spinning planet.

The images

Upper: A Hubble Space Telescope image of the comet about two months before the collision.

Credit: H. Weaver (JHU), T. Smith (Space Telescope Science Institute), STScI, NASA

Lower: Impact sites on Jupiter. On a rocky planet or moon, the collisions would have left significant craters, but since Jupiter is a gassy giant, the consequences were only temporary. They were, nevertheless, dramatic.

Credit: Peter McGregor and Mark Allen, courtesy of Mt. Stromlo Observatories, Research School of Astronomy and Astrophysics, Australian National University

Copyright: Australian National University

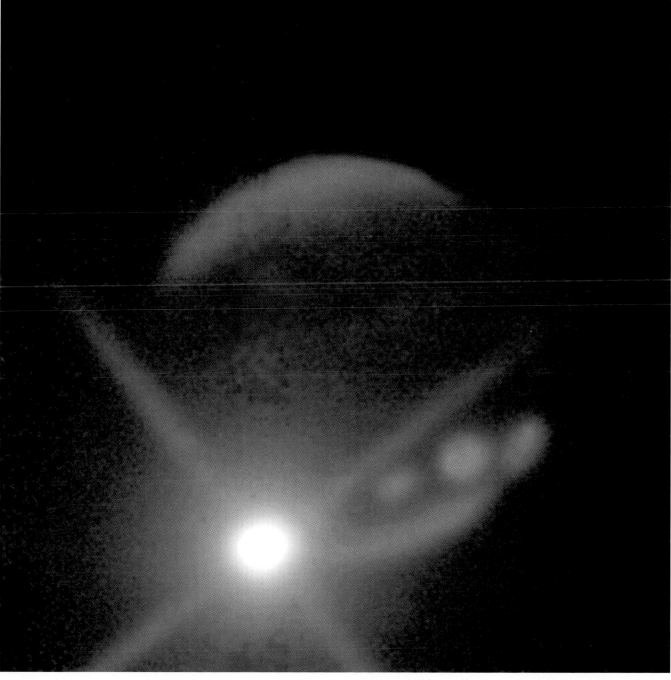

An Innocent Passer-by?

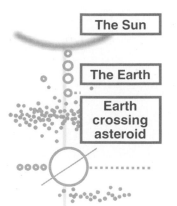

These are three images of one object that is the size of an iceberg. And it is indeed made, at least partly, of ice. But it does not float passively in water. It hangs in space, as do the Earth, the Moon, and the Sun, and it spins just as they do. But it spins faster.

It is called 1998 KY26, and, if you could cling to its small surface, you would move from daylight to darkness and back again in less than 11 minutes.

You would also be on a highly elliptical orbit around the Sun so that you would at times be significantly closer to the Solar System's heart, and at times further away. You might find yourself hurtling scarily close to a larger body such as the Earth or Moon. 1998 KY26 is a Near Earth Asteroid, a passer-by.

In the event of a collision with the Earth, you and your asteroid could be scattered over a wide area. Your dust might form a unique layer around the world, as seems to have happened to the asteroid that made life so difficult for the dinosaurs. 1998 KY26 is smaller than the dinosaur-slayer but big enough to make a dent, and to generate shock waves within the Earth that could trigger violent volcanic activity.

In 1998, it skimmed past the Earth, just 800,000 kilometers away, and was spotted for the first time. Had you been riding on it at the time, you might have feared for your life. You might have feared for everyone's life. But it passed by innocently enough this time.

The object

1998 KY26

1998 KY26 is a fast-spinning and water-rich asteroid, whose pathway crosses that of the Earth. Objects as small as this, only about 30 meters across and faint, are impossible to spot until they get quite close, and there could be about 10 million similar earth-crossing asteroids, each one an accident waiting to happen. There is an awful lot of space out there, however, and the 10 million objects are widely spread. So we live in hope.

The image

This is not a simple set of images, but is constructed using a computer, from observations made using radar as well as visible light. The smooth fruit-like appearance is a result of the modest resolution – no feature smaller than about 3 meters across is clear.

Credit: NASA, JPL, image courtesy of Steve Ostro [Ostro et al., Science 285, 557–559 (1999).]

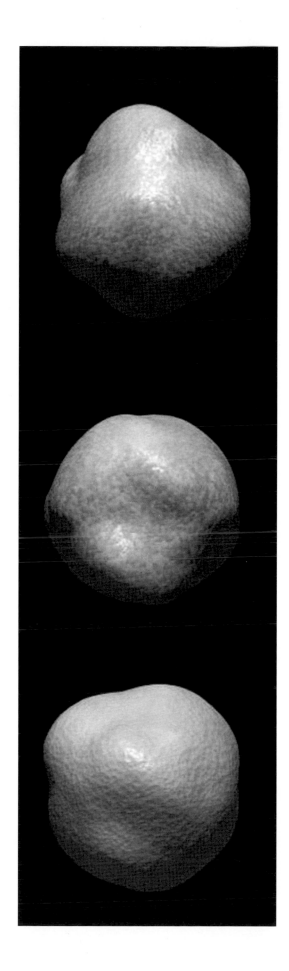

Once Upon a Time, Long Ago

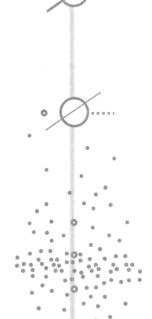

Ganymede is larger than Mercury, and almost half the diameter of the Earth itself. It is the Solar System's largest moon. It is a moon, though, in slumber.

It has a surface of ice mountains, often with long intersecting ridges. There are darker and light regions, all of them ancient. There are craters, but the older ones have not been left exactly as they were formed. Once upon a time, Ganymede did not sleep, but had an active surface. The ice moved.

Flows of ice seem to have shaved and partly buried the oldest features. At times there may have been liquid water, or perhaps slush, on the surface. But all of that was long ago.

There may still be liquid water beneath the icy crust. The only activity on Ganymede's surface now, though, is the rustle of impacts by small grains of dust and very occasionally the thump of a larger body. It has become a place where not much happens.

The object

Ganymede

Ganymede is one of the Galilean moons, visible to a very good human eye, and was first recorded by the Chinese astronomer Gan De in 364 BC but then apparently much ignored until the age of the telescope. Its diameter is 5,262 kilometers, and it has a magnetic field. Like the Earth, it has an iron-rich core, and it has a very thick layer of dirty ice so that about 60% of its volume is frozen water.

The image

Voyager 2, one of the pair of spacecrafts on journeys begun in 1977 and still continuing, produced this image in 1979, when it was only just over 300,000 kilometers from Ganymede. The width of the moonscape seen here is about 1,300 kilometers.

Credit: NASA/JPL – Caltech

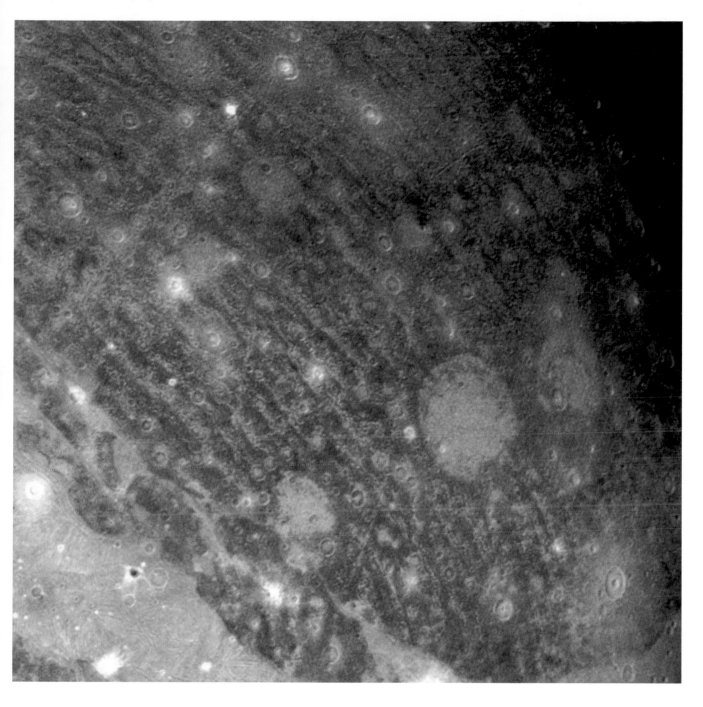

Bigger Than We Are

How many bodies are there in the Solar System? There is the Sun, of course, and eight planets, and there are the smaller bodies: some dwarf planets, comets mostly in the outer reaches (some of them as big as dwarf planets), asteroids, and other debris and dust. The dust is spread out well enough, but there is an awful lot of space and a lot of dust. There are electrons, protons, and ions, but there would be no point in even starting to count them. Numbers don't go that far. And there are moons, which do not have independent orbits around the Sun but travel with larger bodies.

People have, so far, visited just one of all those smaller bodies. The journeys to the Moon by a small and select band of people were a long time ago, but they were journeys with significant cultural impact. For a while at least, people looked outward rather than inward as usual, and maybe made a little more progress toward a better understanding of ourselves as travelers in the Solar System, as small but meaningful parts of something so much bigger than we are.

The object

The Moon

The Moon is quite a lot smaller than the Earth, but not nearly as small in comparison with its host planet as are the moons of Jupiter and Saturn. It is a simplification to say that the Moon orbits the Earth – they spin about their common center of mass, just as two unequal ends of a dumbbell would rotate about a point that is nearer to the heavier end.

The Moon has no significant atmosphere, and is thus exposed to the particles of the solar wind, to radiations such as X-rays from the Sun and from further away, and to a slow sand-blasting by meteors that are mostly but not always small. The latter bombardments have coated its surface with gray dust. There are minerals among the dust but the expense of gathering them would be enormous. The reasons for going to the Moon years ago were scientific and therefore cultural, and political, but not directly economic.

The image

Astronaut Charles "Pete" Conrad took a picture of Alan Bean, and of himself reflected in his colleague's visor, during the Apollo 12 moon landing in November 1969. Bean is collecting samples of dust, but what is most obvious is the level of protection that is needed for any person outside of the Earth's atmosphere. The intensity of the sunlight is just the start of the problems – there are high radiation levels, as well as the vacuum that will burst open an unprotected body.

Credit: Charles Conrad, Apollo 12, NASA

The Sun

The Earth

The Pluto System

Unpossessed

Pluto the god ruled the underworld, and was the god of gold and wealth, the god of death. Pluto, the distant object of the Solar System, came into human consciousness in 1930, and makes its long journeys around the Sun, for the most part, beyond the orbit of Neptune.

For a long time Pluto was thought to be the ninth planet of the Solar System, but it is smaller than seven moons, including the Moon. Then with the discovery of a larger object, Eris, in even-more-distant solar orbit, it became merely a dwarf planet in the human catalog. To the object itself it really doesn't matter how we classify it, however, or indeed what we call it. Though we might pretend that naming is claiming, and delude ourselves that labels confer ownership, its name never leaves the Earth.

Pluto does not travel alone, and more names have been required. Its largest companion is Charon, discovered in 1978 and named for the mythological ferryman of the dead. (Pronunciations of Charon among astronomers are many and various, starting with "sh," with "k" or with the somewhat softer "ch" sound.)

Then, in 2005, appearing for the first time through the optics of the Hubble Space Telescope, came two small and faint moons, a goddess of the night accompanied by a nine-headed monster from the gates of hell, Nix and Hydra. Thus an object once thought to be a solitary dweller in distant darkness has become four, none of them belonging to us.

The objects

Pluto and its moons

Pluto has a diameter of just 2,275 kilometers, and Charon's is just over half of that, at about 1,200 kilometers. The larger of the two is covered with methane and nitrogen ice, but Charon seems to have water ice.

Pluto and Charon both rotate synchronously (that is, they each take the same time for spin around their axes and for orbit around each other) so that they dance forever face-to-face.

In terms of reflected light, Nix and Hydra are about 5,000 times fainter than Pluto, and they orbit the shared center of Pluto and Charon. Beyond that, little is known about them.

The image

The photograph dates from May 2005, and marks the discovery of the two new moons, Nix and Hydra.

Credit: NASA, ESA, H. Weaver (JHU/APL), A. Stern (SwRI) and the HST Pluto Companion Search Team

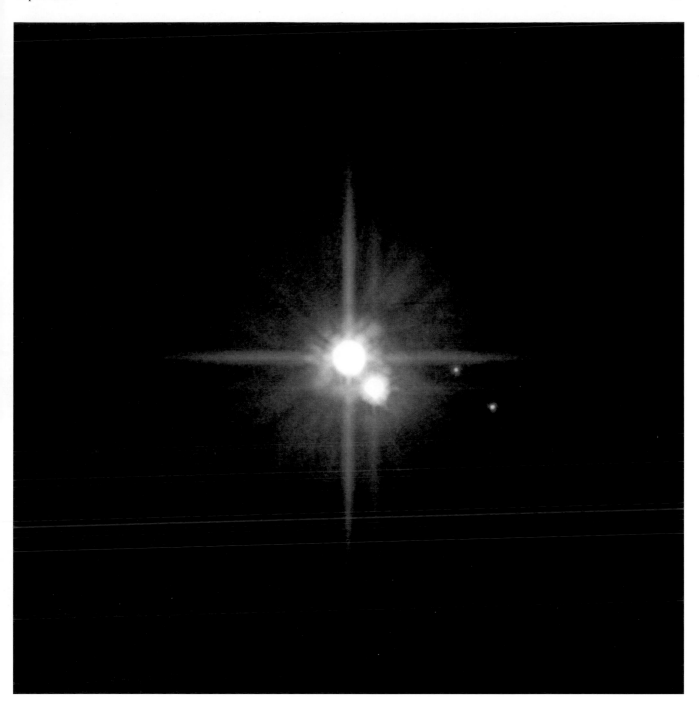

The Sun

The Earth

Saturn

Titan

So Similar, So Different

There are drainage features leading from among the ranges of hills and down toward the lakes. Where in the world could this be? But a prior question should be "which world?" because this image is not of the Earth.

It is Titan, roughly ten times further from us than is the Sun, and yet here it is, mapped in detail and looking much like home as seen from a satellite.

If it were as close to the warmth of the Sun as we are, or if it were closer to Saturn and thus possibly heated by tidal effects, then Titan would be too warm to have an atmosphere at all. As it is, it has gases that are striking in their differences and their similarities when compared with our own protective blanket.

From study of light as perceived by the Cassini spacecraft, detailed chemical analysis is possible. Gases in the atmosphere include methane and ethane, propane, hydrogen cyanide, cyanogens and cyanoacetylene, stirred by the energy of sunlight even from the distant Sun. So, as on Earth, there are weather patterns, and materials in the form of liquids and gases, and as suspended droplets creating the moon's haze. Cycles of evaporation and condensation produce a strange rain that falls on mountains and flows back to the lakes.

The object

Titan

Titan's diameter is 5,150 kilometers, compared with the Earth's 12,750 kilometers. Somewhere on its surface rests the Huygens probe, which left the Earth in 1997 and traveled for seven years in tandem with the Cassini spacecraft, to Saturn and to Titan. The probe gathered data about the Titan atmosphere as it fell to a quiet end on the moon's surface.

It is a very cold atmosphere; the surface temperature is about −180 °C. That is cold enough for the lakes to be methane or ethane, which at temperatures here on Earth are gases. At those very low temperatures, it is very unlikely that anything lives and swims in those lakes. There are, however, volcanic hotspots.

The image

The image is constructed from radar data captured by the Cassini spacecraft's Titan Radar Mapper on April 30, 2006. Titan's atmosphere is murky, and an image such as this would not be possible using visible light, so it was made using reflected radio waves, radar, which can penetrate the smog.

Credit: NASA/JPL

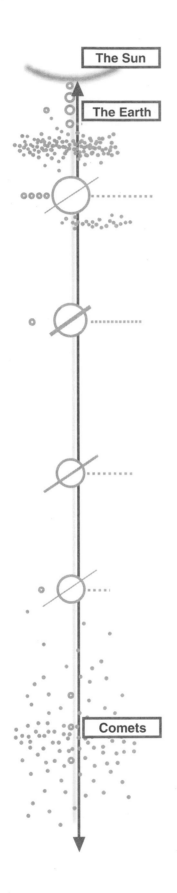

As an Angel

There are, for 24 hours every day, many pairs of eyes looking into the skies. Some are the eyes of professional astronomers while others are making just casual glances. Early in 1997, a comet was visible to billions of people who needed only to look up.

But the keen eyes of two US observers spotted the comet two years earlier, and their names, Hale and Bopp, will forever be associated with its visit. By late 1996, there were some stunning images.

As it approached the inner Solar System, the icy core of comet Hale-Bopp was heated by the Sun and blasted by the solar wind. Material freed from the core filled the space around, and thus the vision grew. From a nucleus of just a few tens or hundreds of kilometers across, a huge enveloping coma of gas and dust developed. This highly pixelated image shows a distance across that cloud of roughly 170,000 kilometers, though it extended very far beyond that.

Local activity on the comet nucleus created jets, and bands of brightness and darkness, giving the appearance here of a human figure with arms outstretched, as a silhouette of an angel. But it is no eternal angel. Each time it comes to the inner Solar System it loses a little more material, and if it avoids a fatal collision with the Sun or with one of the giant planets then, visit by visit, it fades away.

The object

Comet Hale-Bopp

Comet Hale-Bopp was the most easily visible comet for many decades, its coma of dust and gas, when at its largest, being comparable in size to the Sun itself. The estimated time between visits of the comet to the inner Solar System is more than 2,000 years. It has visited before, but not for a very long time. It is now back out beyond the orbit of Uranus and still climbing away from the Sun's gravity, but it will descend again to sprinkle the inner Solar System with its dust when we, too, are dust.

The image

Taken by the Jacobus Kapteyn Telescope at La Palma in the Canary islands, on August 27, 1996, as the comet approached the Inner Solar System.

Credit: Alan Fitzsimmons, Queen's University Belfast/Isaac Newton Group

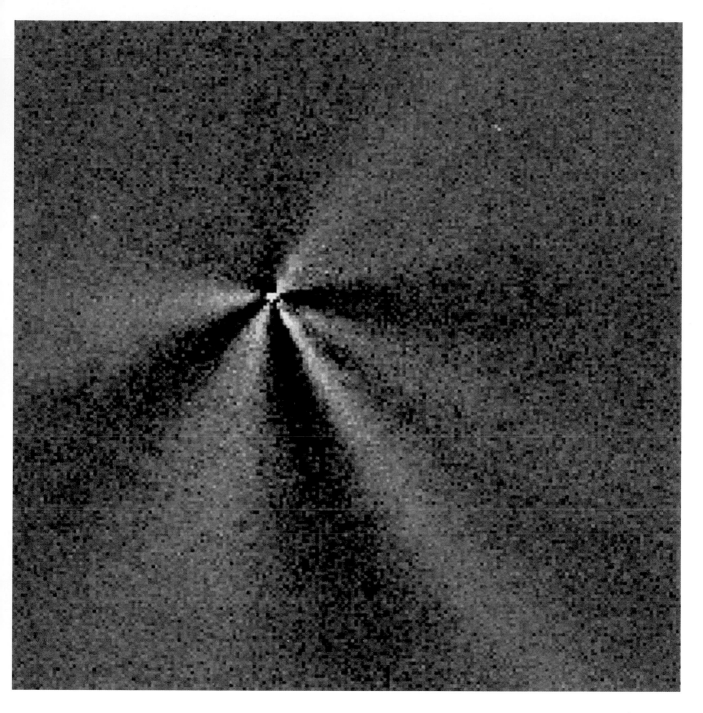

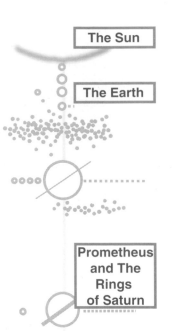

The Sun

The Earth

Prometheus
and The
Rings
of Saturn

Tricks of the Light

The sunlit rings of Saturn seem to unroll like a strip of fabric from the dark edge of the planet. This banded blanket apparently dominates a small moon, Prometheus, though the truth is that the moon is far larger than any of the ice particles of the rings.

The curvature of the blanket at the planet's edge is a trick of the light – the same process as apparently bends a stick part-immersed in water, or distorts the world as seen through a glass tumbler. It is the very trick, indeed, by which the lenses of our eyes make a coherent outside world available to us; a trick that happens whenever light changes speed as it crosses a boundary between one medium and another.

There are boundaries here that are not sharp. The layers of Saturn's atmosphere fade one into another gradually, and likewise into the vacuum above. There are sufficient differences, though, to change the speed of light and thereby change its pathways to create the appearance of curvature where the rings pass behind the planet. Perhaps the most striking point is that this trick of the light that we can see so easily around us is not just a matter of everyday earthbound experience. It is also something that happens far away.

The object

Prometheus and the rings of Saturn

Discovered only in 1980, in images from the Voyager 1 mission, Prometheus is a moon of irregular shape, about 150 kilometers along its longest dimension, taking about 14.5 hours to circle the planet (a distance of about 850,000 kilometers). Prometheus is a "shepherd moon," its gravity working in synchronicity with that of its companion Pandora to prevent particles of Saturn's faint F-ring from dispersing.

The image

This is a visible light image, produced by the Cassini spacecraft on April 28, 2005. Saturn's cloud tops absorb and also reflect sunlight, but the top-most layers are transparent, allowing enough light to reach Cassini's cameras from the rings beyond to show the distorting effect.

Credit: NASA/JPL/Space Science Institute

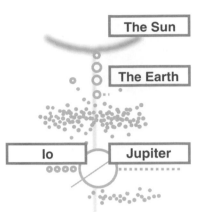

More Than a Match

Earth versus Io:

Which is bigger?

The Earth, easily, with a diameter four times that of Io's.

Which one has the biggest active lava flow in the Solar System?

When it comes to volcanoes, Io is significantly more spectacular.

The image shows the Amarini lava flows with their plumes of sulfur dioxide, a patchwork of color. Much of the flow, hot rivers of rock sometimes hundreds of kilometers long, takes place beneath a crust of old lava. Dark areas of lava have broken through to the surface; snows of sulfur dioxide create brighter regions; sulfur droplets add to the vibrant color.

The Earth is a complex and ever-changing place. It has a coat of many colors – blues and whites, greens and browns. Io's chromatic display is more than a match.

The object

Io

Jupiter's closest moon has mountains as well as huge volcanoes and high cliffs, and compounds of sulfur that give it its colors. Pulled around by its neighbors, Ganymede, Callisto, and Europa, in the awesome presence of Jupiter, Io has an orbit that is less than tidy, and as it rotates around its planet it bounces up and down in an intense gravitational field. The changing distortion of the moon generates the heating effect that results in the violent geothermal activity.

The image

The image, like some others in this book, is a mosaic. The contributing images were taken months apart, in 1999 and 2000, by the Galileo spacecraft. The area is about 500 kilometers long.

Credit: NASA/JPL – Caltech

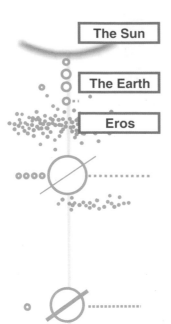

Curiously Normal

It is 33 kilometers long and 13 kilometers across, and it hangs in space far away from any planet. It is the asteroid Eros, a large and irregular stone, one of a horde of such rocks encircling the Sun, further out than Mars but not so far as Jupiter.

Though not nearly large enough for its gravity to hold a cloak of gas molecules above its surface, it hangs on to loose boulders and pools of blue dust as yet unidentified. With these passengers on board, it speeds in one direction and the Sun pulls and turns it from the side, to produce the surprisingly stable condition that is orbital motion. All of the planets and all of the uncounted objects of the Solar System share the same style of pathway. Out there beyond our atmosphere, where there is no resistance to rapid motion, orbit is a curiously normal state of affairs.

In 2000, the NEAR-Shoemaker spacecraft got close enough to Eros to take this picture and itself became a satellite of the asteroid, remaining in orbit for many months before making a soft landing on its surface. The spacecraft's power supplies have long since died, and it now lies inert on the asteroid surface where, we can only suppose, it will long remain.

The object

Eros

The outline of Eros would fit nicely over a large city. Its density is about the same as that of the rock of the Earth's crust. Although most asteroids remain within the confines of the Main Belt, between the orbits of Mars and Jupiter, Eros has a significantly elliptical orbit, so that it crosses the pathway of Mars and at times comes almost as far as the Earth's orbital track.

The image

The NEAR-Shoemaker space probe produced this image in the year 2000, while in Eros' orbit.

Credit: NASA/JPL – Caltech

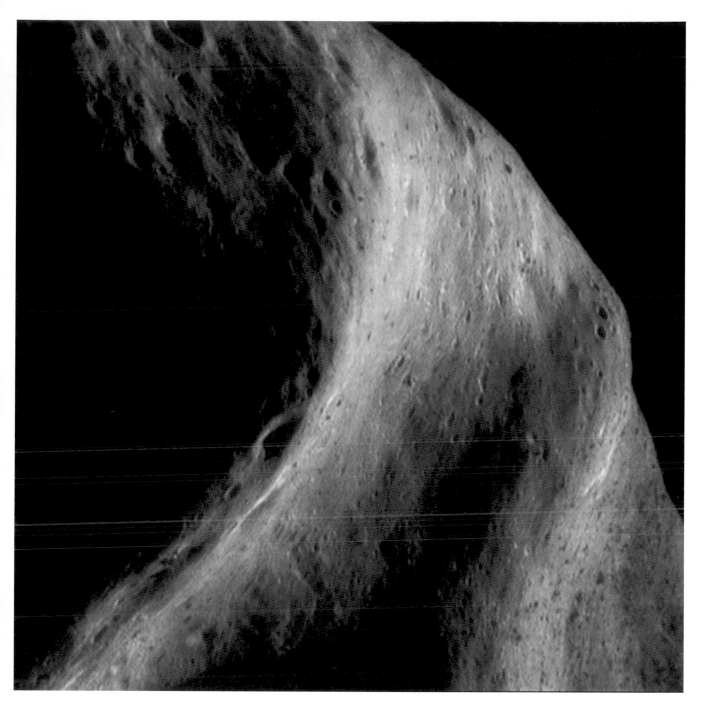

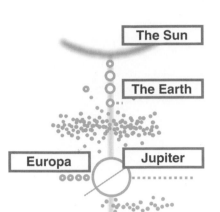

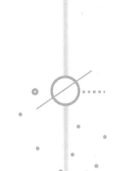

Looking Again

Looking at all of the planets and moons of the Solar System, there are similarities with the Earth – volcanoes here, liquid erosion there, fault lines somewhere else. But it is the diversity that is most striking and most wonderful. Each body has its own very unique story to tell. Learning to read those stories is a great adventure, and by increasing our understanding of other moons and planets, we come to know the Earth better.

Europa, for example, presents a richly threaded tapestry, but reading its story is harder than for the Earth. This is partly for the obvious reason that we do not live on its surface. On our own planet we can witness both the small changes like the crumbling of a riverbank and more dramatic ground-shaking events of crustal fragility. We cannot watch Europa's processes of change nor explore beneath its surface to be able to read its layers so directly.

We can, however, gather information about its outer materials – like its pure water ice, blue in this image, and the salty ice forming the brown strands. Astronomers can make suggestions, hypotheses, about the processes that created these features, and predictions about what they might see by looking again. So they keep on looking, putting their ideas to the ongoing tests of observation.

The object

Europa

Europa is the smallest of Jupiter's Galilean moons. Space missions such as Galileo can probe the inner structure from a distance, for example by detecting magnetic field variations, and so we know that the outer icy layer lies over silicate rocks that are like many on Earth. With less certainty, there are signs of a possible liquid water layer just under the ice, and an inner metallic core.

The image

The image is a mosaic, combining scans of the surface, at different resolutions, made during different orbits by NASA's Galileo spacecraft, almost two years apart – Galileo looked and looked again. The colors are intensified to make the surface features more obvious.

Credit: NASA/JPL - Caltech

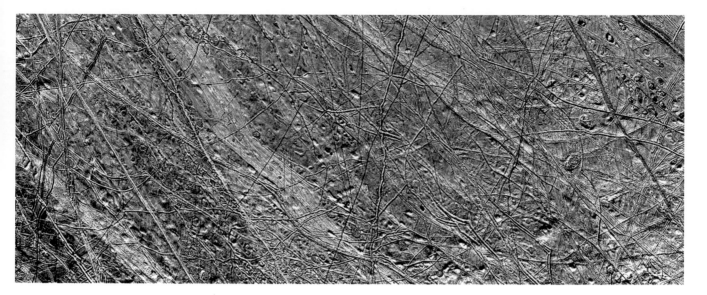

Realities Like This

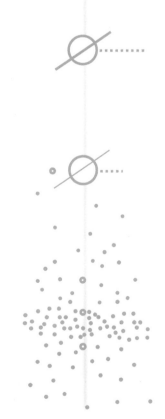

The "ideal" is an aspiration that is beyond reality, but that is embedded in our culture. The fact that it lies outside our physical reach does not prevent us from wondering what perfection might look like.

Is there geometric perfection? Esthetic perfection? Is there a perfect realm, a heaven that contrasts with the imperfect world of the senses? This book does not answer those questions, but is about a different kind of heaven, one that contains Titan and everything else that we might ever see. It is a heaven that contains us and all of our sensed experiences.

The atmosphere of Titan, backlit by the Sun, provides an image that is a matter of the senses, a matter of experience. It is not perfect. It has a beauty that lies not in geometric perfection but in its subtlety, including its delicate asymmetry. It is departure from the ideal that creates interest.

The fading of light into the moon's shadow and the irregularity of the atmospheric layers make a real heaven, not an ideal one; a reality that is as beautiful as it gets. With realities like this, who needs ideals?

The object

Titan

Titan is a large moon, and was thought to be bigger than Jupiter's Ganymede until it was realized that its thick atmosphere made it appear larger than it is. So in terms of size it comes in at number two out of the moons in the Solar System, bigger than planet Mercury.

The image

The Cassini spacecraft produced the image using visible light, on June 2, 2006.
Credit: NASA/JPL/Space Science Institute

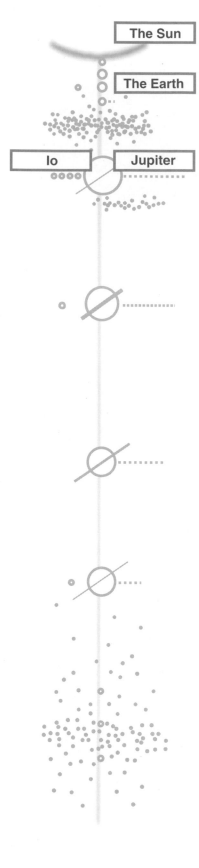

The Sun

The Earth

Io   Jupiter

Above a Turbulent Sky

Io races across the background of Jupiter's raging skies, a small rocky ball above a giant and gassy planet. And race it must. Stable orbit so close to a large host, where gravitational field is intense, is only possible at high speed, and Io finds this stability at 17 kilometers per second, making its long, long journey once around Jupiter in less than two Earth-days.

Io, and Jupiter's three other large moons, Europa, Ganymede, and Callisto, are all visible from where we are with the help of binoculars or small telescopes. And a consequence of their rapid motion is that it is easy to see the change in their relative orientation from one night to the next. Thus it was that Galileo saw, in 1610, that not everything in the heavens moves around the Earth. It was an observation that helped to shatter the geocentric world, casting us out of the center of our own Universe.

The motions of those faraway moons changed technology, changed commerce, changed culture. Their observed orbits promoted revolutions in thought, and the start of adventures that have not ended yet. Space is not some place up above us; we cannot "go into space," since we are already there. We live in space. It is possible to choose to lead flat lives, in two dimensions and not three, to never look up, to dismiss the ongoing adventure of observation, and to choose to be ignorant of the depth of existence. It is possible, but may be unwise.

The object

Io

The diameter of Io is 3,643 kilometers, whereas that of Jupiter is 140,000 kilometers, (Earth's is 12,750 kilometers). Io interacts magnetically with Jupiter, generating huge electric currents and a flow of charged particles into the planet's atmosphere, resulting in Jupiter's auroras.

The image

Taken by the Cassini spacecraft when it flew by the Jupiter system in order to gain energy from the planet's motion and then to travel onward and outward to Saturn orbit.

Credit: NASA/JPL/University of Arizona

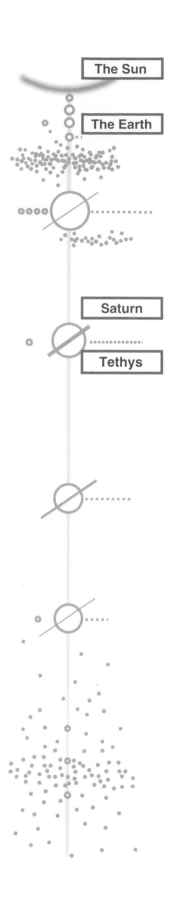

They All Turn out Differently

Take a large drop of water. Very large – about a thousand kilometers across. Freeze it. The surface will crack and ripple. Then bombard it with assorted rocks and snowballs, large and small, for a few billion years. Finally, spray it with fresh crystals from the ice volcanoes of a sister-moon.

You should end up with an ice-ball, quite shiny white in the sunlight, covered in pits and ridges. In fact you'll end up with something that looks an awful lot like Tethys, Saturn's fifth largest moon.

Saturn has an entire family of ice moons – including Mimas, Enceladus, Tethys, Dione, Iapetus, and Rhea – all between 400 and 1,600 kilometers in diameter, and all with density not far from that of water ice. But as with any siblings, their early experiences were not identical, and each one has developed in its own way. Each one is an individual.

Enceladus is the densest of the group and has its ice volcanoes. Dione has enough rock mixed with the ice to have sufficient radioactivity to generate inner heat, just like the Earth. Iapetus has something of a dual personality – with a dark side as well as a bright one. Tethys is made almost entirely of water-ice – like a crystal ball.

As every parent knows, you try to treat them the same, but somehow they all turn out differently.

The object

Tethys

A long trough, a canyon up to 100 kilometers wide and 3 kilometers deep, curves much of the way round the moon, and there is one crater so huge that the impact must have come close to shattering the sphere. There are very many craters, with much superimposition of one on another.

The image

This is a relatively close-up shot, taken on September 5, 2005 by the Cassini spacecraft, from a distance of only 32,000 kilometers.

Credit: NASA/JPL/Space Science Institute

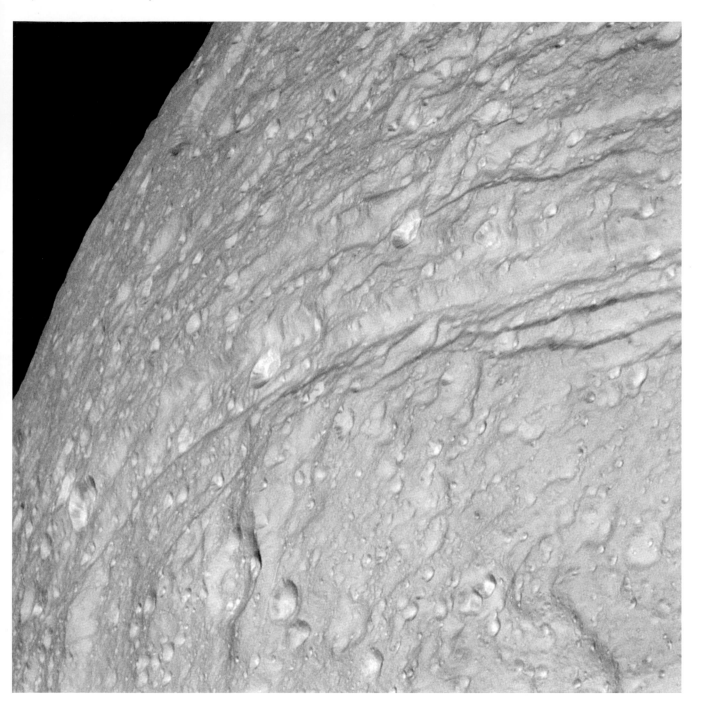

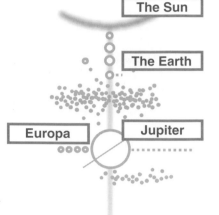

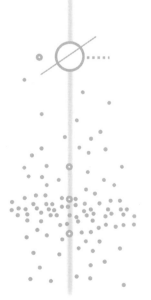

Warhol's Worlds

Andy Warhol's images of Marilyn Monroe and others, in sequences of extravagant color, help to define Western culture of the mid-twentieth century. He did not invent the use of false color, however. Mappers have long used different hues to represent different altitudes of the land. A negative photograph can reverse shades and present an image in a way that does not match the human eye's interpretation of the same scene – the coloring is not true. But it is interesting that the Warhol images predate the use of false color in scientific image making.

As with relief maps, color can represent data, whether it be altitude, gravitational variation, or emission intensity. In medical imaging, satellite imaging of the Earth, or imaging of remote X-ray or radio stars, scientists have found the technique to be a powerful tool in their efforts to see more and so to understand more.

There are two different images here, one in enhanced natural color and one in false color. Each one is then also reversed in color. The effect is striking. To the expert eye, such image manipulation also yields information about the surface of the Europan world.

The object

The Tyre structure, Europa

The circular bruise on Europa's surface is an ancient scar, as yet of uncertain history, 140 kilometers in diameter, laid over with other features – cracks filled with dirty ice, ridges created by the moon's surface movements. There are areas of mineral salts and others that are mostly water ice.

The images

The images were all generated by the Galileo space mission, using three different camera systems, in April 1997. The false color image was taken from a greater distance, and combines data from two imaging systems, while the enhanced natural color image was taken at much closer distance.

Credit: NASA/JPL – Caltech

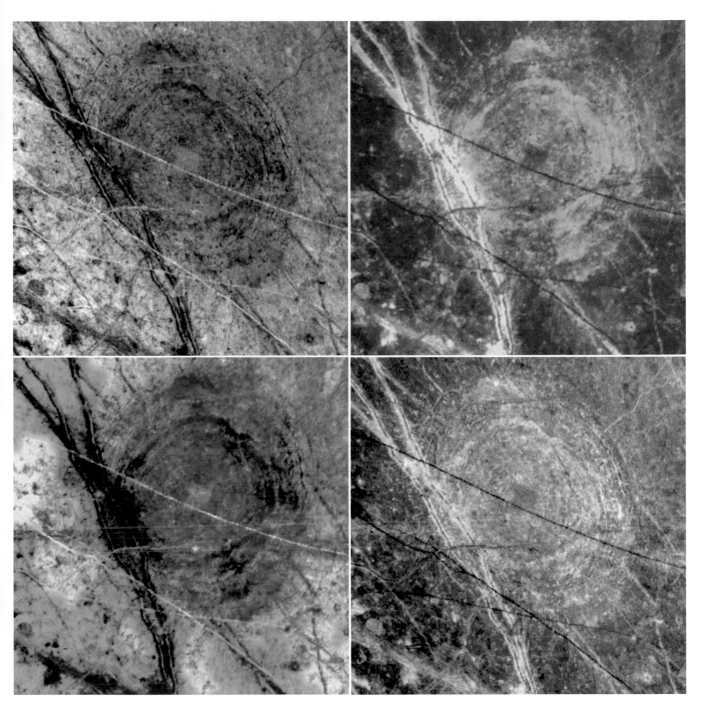

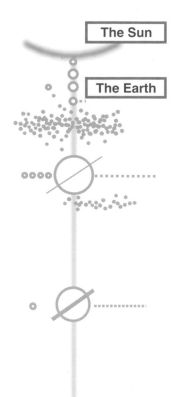

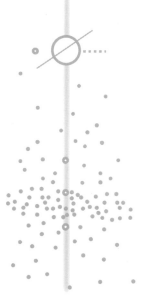

An Edge of Darkness

The Earth's terminator sweeps over us as certainly as death's scythe, but more often. Twice every day the circular edge of darkness creates a twilight zone for each of us, a boundary between sunlit day and the hours of the night. Miranda, a small moon of Uranus, has a terminator, too, of which this image shows just a small portion.

Miranda itself lives on an edge. The strength of its gravity holds it together, but doesn't pull hard enough to flatten large and irregular shapes. Right at the terminator in this image, right at the edge of darkness, is a cliff that is higher than the sides of the Grand Canyon.

The Voyager 2 spacecraft only passed by Miranda because it needed to, in 1986, in the process of stealing a little energy from Uranus for its ongoing journey out to Neptune and beyond. Not much was expected – perhaps a static little moon, boring even. But the shapes and the processes that may or may not have produced them turned out to be rich material for thought.

Unfortunately, more evidence about Miranda's processes will be hard to come by, perhaps even impossible. The close-up Voyager pictures are all that anybody has to work with. Going out to Uranus' orbit for another quick look is not an option, so it might be that Miranda's history will remain hidden from us, forever in darkness.

The object

Miranda

Miranda is just 470 kilometers across and has a density that is not much more than that of ice, of which it is mainly made, with some silicate rock and organic compounds. The surface has craters and many other features, indicating a rich geological history.

The image

The Voyager 2 spacecraft flew close to Miranda, just 36,000 kilometers away, to make this high-resolution picture. That was in 1986, and provided a brief but detailed glimpse of the moon. Voyager is still on its journey, now out beyond the planets, and no mission is planned that will provide another such image of Miranda.

Credit: NASA/JPL – Caltech

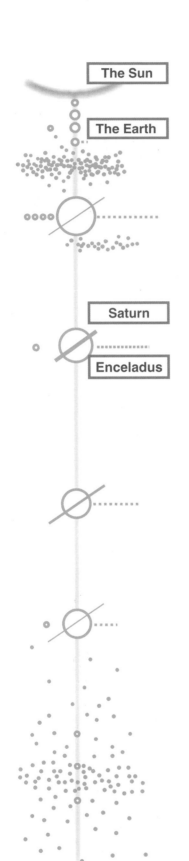

The Sun

The Earth

Saturn

Enceladus

No Moon Is an Island

Enceladus is not an isolated sphere, existing without connection to others. Neither is it an island, existing without connection to others. Not only is its icy globe bound to Saturn, but it shuffles in time in a complex synchronicity with other moons. No moon is an island.

While other moons appear quite dead, Enceladus can be seen in other images scattering ice into the sky, leaving a trail of crystals all around its own orbital pathway, contributing to Saturn's ring system. Not only that, but part of the moon can be seen to be well cratered while elsewhere it is instead crossed by long, long cracks and crinkles. The absence of craters in some regions suggests recent and perhaps continuing surface activity – craters here are soon wiped clean by other ongoing changes. Showers of ice crystals may well be providing fresh "snowfall."

It is not difficult to see the relative ages of superimposed craters. The largest crater in the upper right portion of the image has a smaller crater that disrupts its rim and so must be more recent. Above that is another large but sharper crater, less powdered with dust, lying on top of a smaller and older one.

The effect is the uniquely textured surface, looking like the froth of a coffee or on a glass of beer. But bigger, harder, colder, and out of direct reach. It is not entirely out of reach of human observation and understanding, however. With us to admire it, Enceladus is not alone.

The object

Enceladus

Enceladus is not large, about 500 kilometers across, the diameter of a smallish country. It lies within Saturn's E-ring, and it may be the source of the ring, through the cold volcanoes that throw ice flakes into Saturn orbit.

The image

The image is built from several infrared, visible and ultraviolet scans made by the robot spacecraft Cassini when close to Enceladus on two separate dates, March 9 and July 14, 2005. The blue color is false, but serves the purpose of emphasizing the long linear features.

Credit: NASA/JPL/Space Science Institute

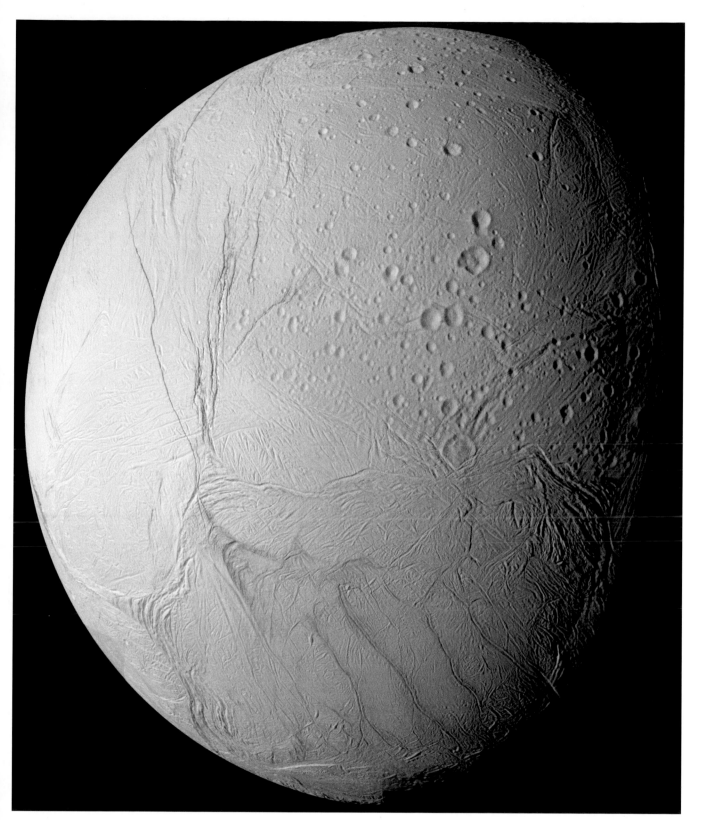

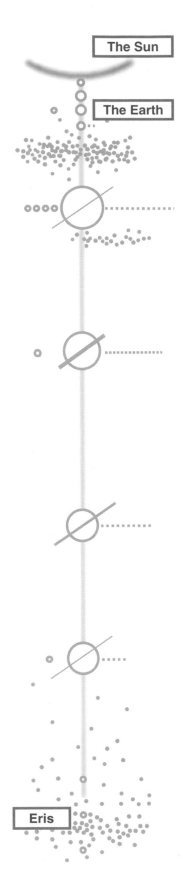

The Sun

The Earth

Eris

What Shall We Call Her?

It's bigger than Pluto, and for a small part of its 560 year orbit around the Sun is actually closer to us than Pluto is. It has pretty much as much right to be called a planet. It was just discovered 70 years later, that's all. One consequence of the discovery was that Pluto was officially demoted to be merely a "dwarf planet," along with the new object (and with the large asteroid Ceres).

The discoverers, in 2005, using images made in 2003, were Mike Brown, Chad Trujillo, and David Rabinowitz. Mike Brown had just gained a daughter, as well as a new dwarf planet, and in both cases some thought was needed about names. Respectively, the names given were Lilah and, temporarily, 2003 UB313. Most girls would prefer to be called Lilah than 2003 UB313.

A baby's name is pretty much a matter for the parents, but for new Solar System objects like 2003 UB313, the discoverers propose names and then it is an international committee who make the final choice, after mining deep into ancient mythologies. They chose Eris (pronounced eeris), the name of a mischievous goddess who created conflict by using the simple ploy of making people believe that their opinions were the only correct ones. Clearly she is still at work in the world. Given the strengths of opinions among astronomers over the status of Pluto and Eris as full planets or dwarf planets, it was a well-chosen name.

The object

Eris

Light takes almost a day to reach Eris from the Sun and then return, after reflection, to us here on Earth. That great distance provides the reason why the discovery took so long. (Eris has an elliptical orbit, however, and when the current millennium is a quarter of the way through, it will be at its nearest point to the inner Solar System and to Earth, much closer than it is now.) It seems to be a bright object and it may be that it has extreme seasons, lasting hundreds of Earth years, so that when it is furthest from the Sun it is covered in frozen methane and nitrogen, and when it is closest these materials sublime into a seasonal atmosphere.

The image

It was the Hubble Space Telescope that made this image of 2003 UB313, showing it to be 1.5 picture elements (pixels) across, equating to the diameter of 2,400 ± 100 kilometers. Further work with the Earth-surface Keck Observatory, using special techniques to compensate for the distorting effects of the atmosphere, shows that Eris has a moon in close attendance, a companion in mischief, named Dysnomia after a goddess of lawlessness.

Credit: NASA, ESA, and M.Brown (California Institute of Technology)

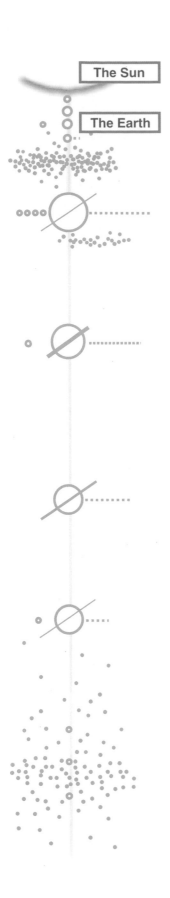

The Sun

The Earth

The Wrong Place at the Wrong Time

Intelligence, it seems, can arise on Earth. It's a fanciful idea, but imagine that there were sophisticated dinosaur technologies, advanced trading systems and global collaboration that surpass our own. Yes, it's fanciful, but no more so than the self-worshipping notion that the dinosaurs were some sort of evolutionary blunder and only now have we arrived as the supreme achievement of the Universe.

Even if there were, 65 million years ago, a species as clever as we suppose that we are, it wouldn't have helped them much. Some natural disasters are going to get you anyway.

If your home planet is going to be struck by an asteroid, and supervolcanoes are going to erupt, there really isn't a lot that you can do. You may not, for a start, get much warning, either of bombardment from above or catastrophe from below. You might as well close your eyes and wait for the bang and for the bleak global winter that follows.

Which played the bigger part in wiping out the dinosaurs, asteroid or supervolcanoes, is still a matter of scientific debate, and the search for more evidence continues. But there was a big impact 65 million years ago. The evidence lies under the coast of the Yucatan peninsula in Mexico, a crater roughly 200 kilometers across, buried beneath the sediment of all those years.

There was no design flaw in the dinosaurs. They were just in the wrong place at the wrong time.

The object

The Chicxulub crater and its asteroid

The Earth has craters, but compared with those on most of the moons of the Solar System, they do not last long, just hundreds of millions of years at best, and even then they are soon much eroded or buried by the sands of time. We can find them, though, and so it was for the Mexican crater, named after the village of Chicxulub.

The asteroid that produced the crater must have been in the region of 10 kilometers across, perhaps the largest object to crash down in the last billion years. At the same age as the crater, is a very thin layer of unusual material, rich in iridium, all around the world, quite probably the dust of the dinosaurs' doom.

The image

Surveys commissioned by the Mexican National Petroleum Company revealed the crater structure in 1990. It is shown here in false color, in an image constructed not from light but from gravitational and magnetic data.

Credit: V L Sharpton – University of Alaska-Fairbanks and the Lunar and Planetary Institute

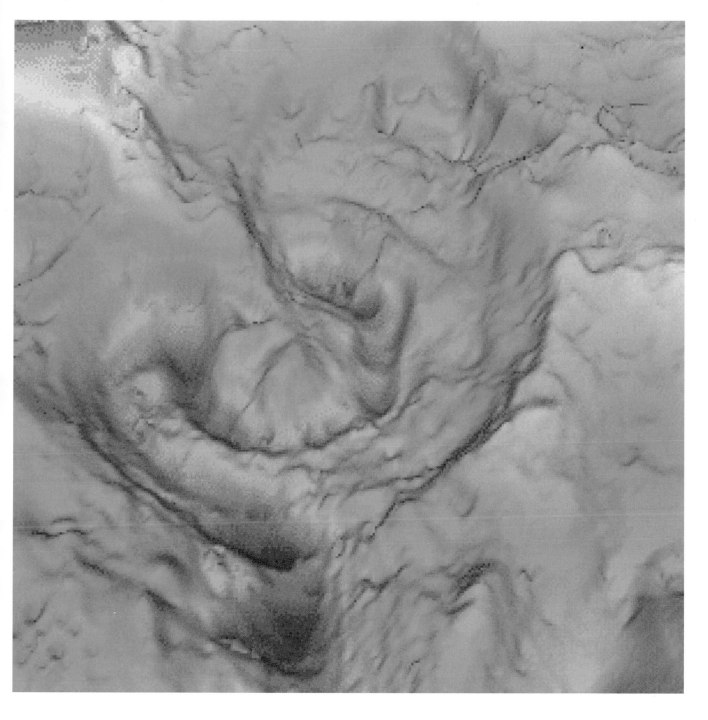

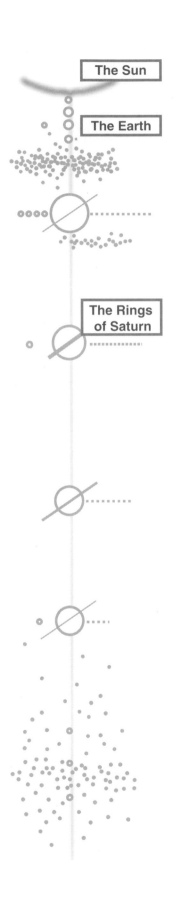

The Sun

The Earth

The Rings
of Saturn

Heaven's Harpstrings...

More than 150 years ago, James Clerk Maxwell developed a mathematical prediction that the disk around Saturn could not be a rigid solid. A solid disk would fall apart, he said, thanks to gravity and its own orbital motion. Observations that could check his prediction, and support it, came much later; firstly from the spectrum of reflected light from the rings (about fifty years later), then from the Voyager spacecraft (more than a hundred years later), and recently by close-up imaging by the Cassini mission. The rings are grainy bands of ice, with disturbances trickling through them in time with the motions of nearby moons.

These disturbances flow through the rings much as ripples pass across water or vibrations move along the strings of a harp. But they are not blown by the wind; there is no wind to blow. They are not plucked by mighty hands, unless you count the passing of moons such as Prometheus and Mimas.

Some of these vibrations travel inward toward the planet, and others move outward, so that they cross over, harmonized with the moons' orbits. The mathematical interpretations of these ice waves follow a long line of descent from the ancient harmonies of Pythagoras, through the genius of James Clerk Maxwell to the present day, and indeed the challenge of mathematical analysis of the observations of the Cassini spacecraft is yielding new methods, new ways of predicting nature's ways.

The object

Saturn's A ring

Saturn's A ring is one of Saturn's most prominent rings, being more than 14,000 kilometers in width, yet only a matter of tens of meters thick. The surprising stability of such a structure seems to be partly the result of the gravitational effects of moons that accompany the ring; it is a puzzle to which mathematicians are seeking solutions even now.

The image

The picture, showing part of Saturn's A ring, was made by Cassini's narrow-angle camera at the time that the spacecraft was entering Saturn's orbit for the first time, in the summer of 2004.

Credit: NASA/JPL/Space Science Institute

...And the Gates of Hell

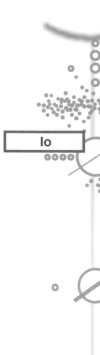

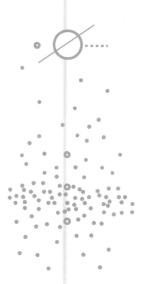

Hell, traditionally, is below. Certainly there is plenty of fieriness deep beneath our feet, generated by the planet's own radioactivity, and sometimes manifesting itself at the surface with dire consequences.

But the Earth is not alone in the Solar System in having a heated structure beneath a thin shell, nor in having that shell broken, violently, from time to time and from place to place by the inner activity. Unlike the heart of the Earth, though, Io's interior is not self-heating. It is a moon that moves closer or further from its planet, pulled out of a more regular and more circular orbit by its fellow-moons, Europa, Callisto, and Ganymede. Thus it is warmed by the huge tidal effects of its regular kneading by Jupiter's gravity. The molten layers of our own underworld are hellish, but so too are the volcanoes of Io, high in the heavens.

The object

The volcanoes of Io

Like the Earth, Io has few craters. Bodies that have surfaces that change relatively quickly wipe their own faces clean too quickly for craters to survive for long. In Io's case it is its volcanoes that do the wiping.

Tvashtar Catena is a chain of active volcanoes, complete with lava flows sometimes hundreds of kilometers long, on Io's surface. The bowls of the volcanoes can be several kilometers deep, and there are lakes of molten sulfur. Huge plumes of material, sulfur or sulfur dioxide, reach far into the Io sky. Visits by the two Voyager spacecraft and much later by the Galileo mission show that the patterns change over quite small timescales – matters of months. Io, being so much further from the Sun than we are, has a low average temperature, much lower than the Earth's poles in winter, but the volcanic hotspots can reach 2,000 °C. Hell indeed.

The image

The image is a combination made with infrared and violet filters in order to emphasize temperature differences so that the very hot main lava flow stands out strongly. The pictured area is about 250 kilometers across, and the visible bright lava flow is about 50 kilometers long.

Credit: NASA/JPL – Caltech

Simple Rings

All of the outer planets have ring systems. Saturn's are by far the most flamboyant. Jupiter doesn't flash its rings around for all to see quite so obviously.

From the darkness of the far side of Jupiter, the Galileo spacecraft saw sunlight scattered by the upper limits of Jupiter's atmosphere creating a fine arc of light. And reaching outward was another fine line – Jupiter's main ring, seen almost "edge-on." The ring is a tenuous strand of fine particles, misty-edged, and in this image its simplicity is its beauty.

The main ring is associated with the orbits of Adrastea and Metis, two of the planet's small inner moons. So it seems likely that the dust is thrown up from the moons by meteor impacts, which are powerful, thanks to Jupiter's accelerating effect. Simple beauty has dramatic origins.

The object

The rings of Jupiter

Jupiter's ring system is made of dust and not mostly ice like the rings of Saturn. In this image, only the main ring is visible, seemingly slender but in fact its width is about the same as the radius of the Earth. There is also an inner halo, and much fainter outer rings. While the other rings are made of grains of matter thrown into independent orbit by violent meteor impacts with moons, the very ghostly outermost ring may simply be captured interplanetary dust.

The image

The rings, like the edge of Jupiter, become visible by the scattering of sunlight. With no sunlight, the rings vanish, and here the near part of the ring close to Jupiter is in its shadow. This is a single-color image, taken using a filter by a camera aboard the Galileo spacecraft in 1996.

Credit: NASA/JPL – Caltech

The Sun

The Earth

Europa   Jupiter

Crime Scene Investigation

Call in the crime scene investigators. There is a bullet wound on the face of Europa.

- Central wound: 26 kilometers in diameter, characterized by a circular ridge around a depression with a central peak.
- Stain around wound: 40 kilometers in diameter, probably due to material from below the surface.
- White radiating pattern: Some rays extend for much more than 100 kilometers in many directions, superimposed on the surface below, suggesting that this was a recent crime.
- Suspects: Asteroid or comet on the loose.
- Required action: Hope it doesn't happen to you.

The objects

Europa and the Pwyll crater

Europa is the smallest of Jupiter's four large Galilean moons, but possibly the most interesting. Observations by the Galileo spacecraft strongly support the idea that there is salty water beneath Europa's icy crust. In similar conditions on Earth we should not be surprised to find life.

The Pwyll crater is a younger feature on older terrain of cracked and ridged ice. Pwyll, meaning "caution," is the name of a hero in the Mabinogion, an ancient book of Welsh mythology. (The double LL is pronounced by curling the tongue back and blowing air out either side. The anglicized version of the name is Powell.)

The image

The image is a mosaic created by the Galileo spacecraft in April 1997.

Credit: NASA/JPL – Caltech

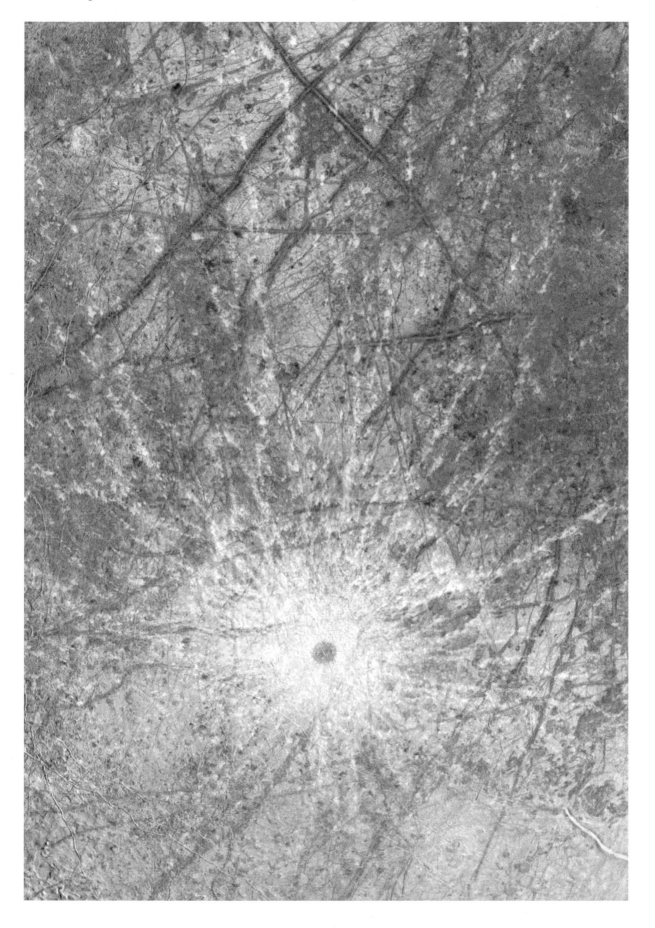

The Dust Left Behind

It isn't always necessary to send missions into space in order to observe smaller bodies of the Solar System, nor to use sophisticated ground-based observations. Some of them come to us, and we have only to look upward to see the spectacle.

Comet Swift-Tuttle swings in around the Sun and out again beyond all of the planets, time after time. It drifts across our skies for several months, once every 130 years or so, and will do so as human generations come and go, until there is nothing left of it. The comet is a ball of dirty ice, perhaps 10 kilometers across, and every time it joins us in the warmth of the Sun it loses a little more material, leaving a trail of dust. Every year, in August, the Earth travels through a past trail of Swift-Tuttle, through the dust it left behind. And we are treated to a fine spectacle, the Perseid meteor shower.

From the ground, with casual observations by eyes alone, it is hard to judge distance to objects in the sky. It is not obvious that these shooting stars are so close to us, just tens of kilometers above our heads, so much nearer and so much smaller than the fixed stars. The largest ones can be, for the second or so after they come in from the void and before they burn away in their strange new airy surroundings, the brightest objects in the night – veritable fireballs. All of them remind us of our place, protected only by a thin atmosphere from the enormity of a greater reality.

The object

The Perseid shower

Perseus is one of the ancient constellations, patterns of visible stars that have no special relationship with each other except, from where we are, a fixed orientation. As such, the constellations are convenient "landmarks" in the sky – spacemarks, perhaps. It just happens that the shower of meteors appears to come from the direction of Perseus, as the Earth passes through the comet dust left in the trail of Swift-Tuttle, once a year.

The image

The photograph is not a simple one, but is constructed by superimposing many images with the same background, in order to capture as many meteor trails as possible. It was produced by Fred Bruenjes during the annual Perseid meteor show, on the night of August 11–12, 2005, and shows the meteors apparently radiating from a point. In fact their motions are parallel to each other, and this is just a perspective effect.

Credit: Fred Bruenjes

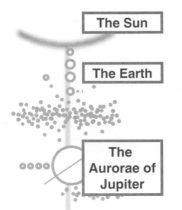

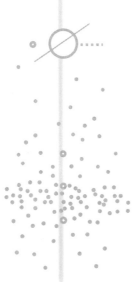

Skies on Fire

There are an awful lot of bodies in the Solar System, from the Sun itself to specks of dust. Among them there are submicroscopic, subatomic particles that leave dramatic traces even though individually they are profoundly undetectable from here on Earth.

The Earth has its glowing auroras, its Northern Lights and its Southern Lights, created by charged particles that stream out from the Sun, reaching the Earth at speeds of several hundred kilometers per second. The Earth's magnetism twists and twirls their pathways so that they rain down toward the poles, producing a phenomenon that is akin to the flickering of a flame. It is ionization, caused by the extreme energies of the incoming particles rather than by the heat of fire. The air itself blazes, in shimmering curtains of light, in response to the solar bombardment.

Jupiter has auroras, too. And the solar wind again plays a part, though the sources of much of the falling dust are the volcanoes of Io. As on Earth, the tiny particles tumble down magnetic pathways, and the atmosphere itself becomes light.

The object

Auroras

Auroras are not objects but transient effects. They are emissions of light, related to events taking place at deep levels within matter. The energy for the changes comes from bombardment by charged particles, whether ions from Io or the electrons and protons of the solar wind. The particles rain down on the polar regions of magnetic planets, like Earth and Jupiter.

The image

This is an image from the Hubble Space Telescope, taken in November 1998 by detecting ultraviolet emissions. The planet reflects visible sunlight, and in ordinary images, the auroras do not stand out well against this light. But it reflects little ultraviolet radiation, and so in this image the rest of the planet is dark compared with the bright auroras.

Credit: NASA/ESA, John Clarke (University of Michigan)

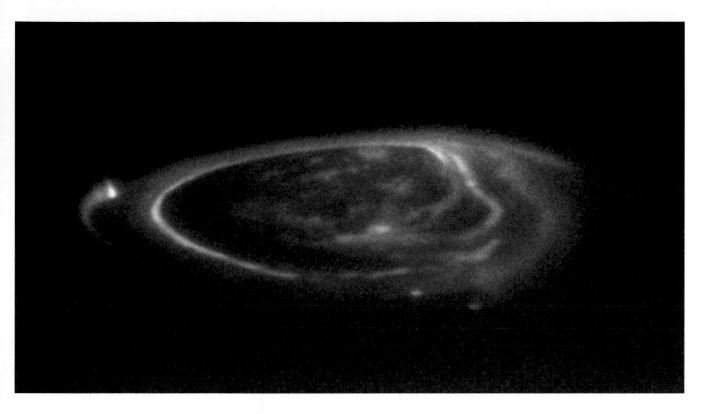

Looking on the Bright Side

On Titan, every day is cloudy. It is dull too, illuminated only by a very distant Sun. On the bright side, though, the clouds are a cheerful orange color.

Here they are seen in black and white only, but that perhaps emphasizes their subtlety. It seems that Titan's atmosphere, like the Earth's, is a place of continuous and complex change, of mists and rain. They are, though, unfamiliar mists and different kinds of rain.

The comparison with the Earth helps us to see the home planet anew, as one world among many. There is huge diversity in all of the worlds that we can now see in detail in our own Solar System, and who knows what lies in all of the systems of so many more stars. Very many more kinds of rain, perhaps – rains of every flavor and every color.

Will future generations of people ever experience such rains? One view might be that it would be a shame if they do not, if human awareness were forever limited to a small corner of the Universe, when there is so much out there as yet unknown. Alternatively, perhaps, we should be more easily satisfied, and know our place. Whichever is your personal view, the important knowledge is that we are one small part of a Universe that has orange clouds as well as gray.

The object

Titan

Titan is never closer than a billion kilometers from us, a seven year journey for the Cassini spacecraft, but certainly an interesting world – bigger than the planet Mercury, with atmospheric pressure greater than on Earth, and with a dynamic atmosphere that reminds us of our own.

The image

The image was made on St. Valentine's Day, 2005, using infrared light scattered from the rays of the Sun.

Credit: NASA/JPL/Space Science Institute

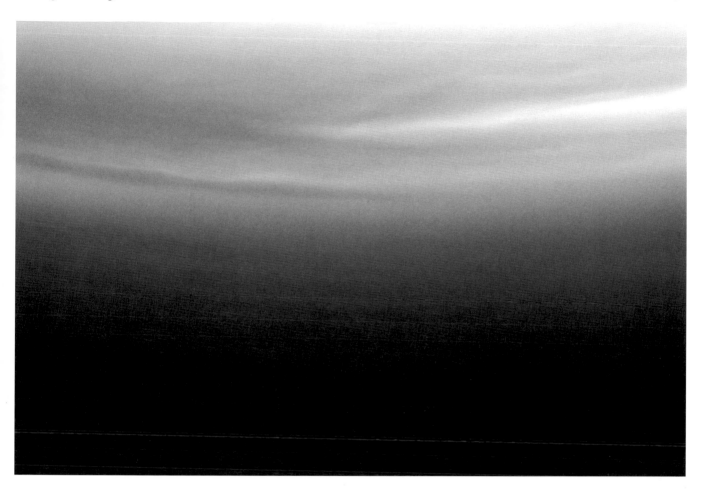

Fragments of Truth

The thin rings of Saturn form a straight line across the background of the planet while a small moon, Tethys, hangs below, in one of the spectacular images beamed back to Earth from the Cassini spacecraft. The result is a picture of harmony, and of apparent stillness. But it is the camera that has introduced the stillness, and has frozen the scene. The camera might not quite lie, but it presents a very limited truth. Time and change are part of nature's fabric, inseparable from the whole. In nature there are no frozen moments.

Change involves the battering of Tethys by smaller bodies, making craters on its moonscape that overlie the results of more ancient events that took place perhaps billions of years ago. Change includes the rapid orbit of Tethys around Saturn, repeating again and again and again. Change, though denied by its product, is essential for the camera's process in snapping its little moments.

The human eye does not work with still frames but with continuity. Yet we somehow accept still images as normal, and are inclined thereby to forget our own entanglement with reality's flow. It is a foolish lapse, and the images of this book and others, however beautiful and however significant, should carry a warning: These are but fragments of truth, and can mislead the unwary.

The objects

Tethys and the rings of Saturn

Tethys is less than a tenth of the size of the Earth in diameter, and with a much lower density, suggesting that much of its material is frozen water. It has some large craters, and a huge crack that wraps much of the way around it, exactly as if expansion of its interior had pushed open a thin outer layer.

The rings are slender disks. Even at their thickest, they are barely more than a kilometer from one side to the other. Here, caught in edge-on view, that nature is revealed.

The image

From 2.5 million kilometers from Saturn, the Cassini narrow-angle camera generated this image on December 3, 2005. The location of the camera was such that it saw sunlight fall on just a crescent of the moon and also Saturn itself, but clever cropping of the image hides the planet's crescent appearance, emphasizing that a camera offers the truth that is spatially as well as temporally limited.

Credit: NASA/JPL/Space Science Institute

The Sun

The Earth

Tethys and
the Rings
of Saturn

All Her Thoughts

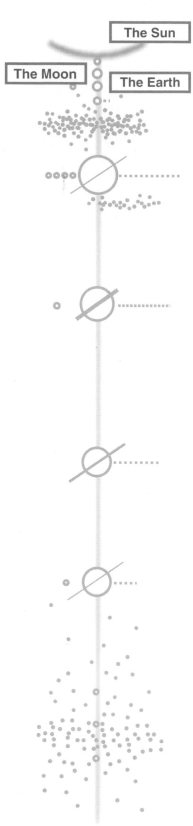

Imagine a scientist at work, trapped deep inside a building and receiving this image on her computer screen. She would, almost certainly, be struck by its beauty, and might want to think about what it shows, with a fresh mind, as if seeing such a sight for the first time.

There appears to be a moon or planet with its full disk flooded by starlight. The edges of the object are not quite sharp, either because the body itself is surrounded by a deep layer of gases, or because the image has been produced from within the atmosphere of another body so that light becomes partially scattered before reaching the camera. The latter seems likely, because the surrounding structures extend too far to be associated directly with the shining body, and seem to be part of some foreground.

Such an analysis could lead to a conclusion that the foreground body has a complex and dynamic atmosphere; one that contains clouds and must therefore contain material that changes state between gas and droplets of liquid. To a fresh mind, her conclusion is stunning. But perhaps not as stunning as her existence, with her sense of beauty and all her thoughts about the Universe of which she is part.

The object

The Moon

The Earth's Moon is one of the larger "smaller bodies" of the Solar System, being the fifth largest moon (after three of Jupiter's and one of Saturn's). At all times, as for other moons and planets, half is illuminated while the other half remains in darkness. We see greater or smaller portions of the sunlit half, as each month proceeds. (It is in fact possible to see soft Earthshine reflected back from the dark side of a crescent Moon.) Though some storytellers will claim otherwise, you will never see a thin crescent Moon at midnight – its direction in the sky must be not far from that of the Sun if most of its dark side is facing us.

The image

The photograph was taken by Fekete Csaba in Hungary in the winter of 2004.

Credit: Fekete Csaba

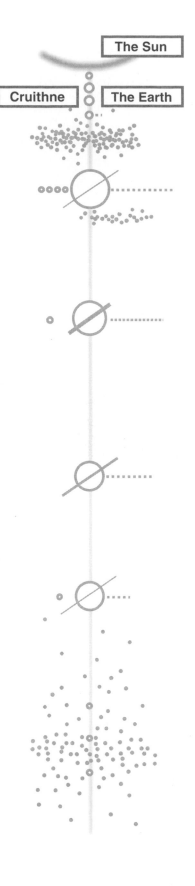

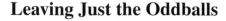

Leaving Just the Oddballs

The Moon exists in close gravitational association with the Earth. There are other bodies, also traveling around the Sun in the same length of time as our planet and at much the same average distance from the Sun. These move in complex patterns that can in no way be called orbits of the Earth. They are not our moons but our companions.

The first to be discovered, in 1986, now has the name Cruithne (pronounced roughly as crooeenye), after a king from ancient Scottish tales.

Their motions are bizarre, but synchronized with our own. They pull toward the Earth and fall away again, so that they seemingly bounce around the Sun, taking an Earth-year nevertheless for one solar orbit, just as we do.

Cruithne and the others are like asteroids, small lumps of rock, and collisions with the Earth would seem more than possible. They don't happen, however, mostly because the orbits of these companions are strongly tilted relative to our own. There may once have been others sharing the Earth's orbital timetable. But any such companions that were going to collide did so long ago, leaving just the oddballs.

The object

Cruithne

The orbit of Cruithne is in 1:1 resonance with that of the Earth, which is the technical way of saying that, on average, it takes the same time to orbit the Sun. An alternative name for such an object, rather than "companion," is "co-orbital asteroid."

Cruithne is only about 5 kilometers across. In the process of trading energy with the Earth due to their mutual gravitational pull, the little companion experiences an orbital adjustment of about half a million kilometers while the planet's pathway experiences a matching change of about 13 millimeters.

The image

Only a movie can really show the complexity of the relative motions of the Earth and Cruithne around the Sun. This is a still from such a movie, showing that Cruithne orbits the Sun and not the Earth, but shares the Earth's solar orbit in an interesting way.

Credit: Paul Wiegert, The University of Western Ontario

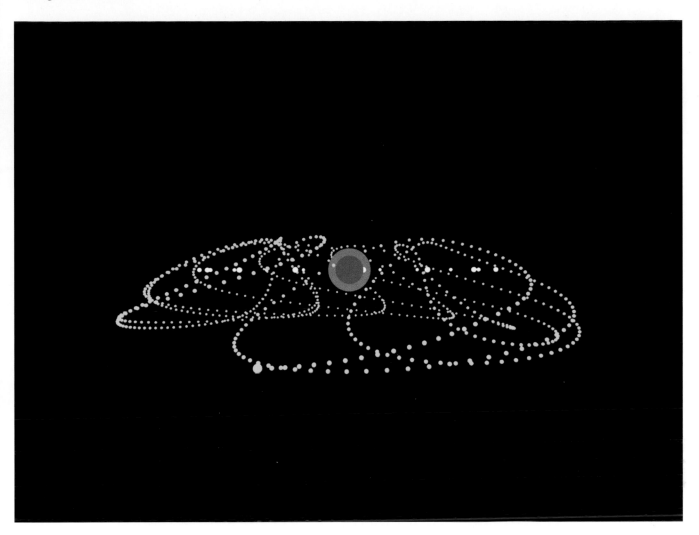

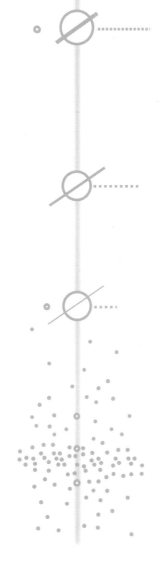

The Sun

The Earth

Europa   Jupiter

Epic Stories

It is an epic story, full of colorful characters, dangerous voyages, conflict and tragedy, and a big surprise. It is the story of how inhabitants of a planet became aware of what goes on beneath their feet.

The central figure is Alfred Wegener, a meteorologist of the early twentieth century, who dared to speak out, speculatively, on matters of geology. He saw patterns in the continents, key similarities and differences in rocks and fossils, and spoke of the possibility of continental drift. For this he was ridiculed. Long after his tragic death on a Greenland expedition, it turned out that he was right. The discovery of sea-floor spreading showed that the continents do indeed move around the globe – a discovery that should be a big enough surprise for anybody.

Now there are whole new worlds, dozens of them, like Europa, all demanding science at its most creative. Gathering data directly from nature and matching it to coherent theory is no sterile activity. It requires imagination, insight, originality. It requires debate and ruthless rejection of ideas that do not stand the test of observation after observation. For every new world there are new challenges of explanation. The surfaces of Europa and Ganymede, Titan and Iapetus, Mimas, Triton, Rhea, and Enceladus – each one is a new epic story.

The object

The icy crust of Europa

Like the Earth, Europa has a crust that is seemingly solid but is a crust that moves. It seems in this image that sections of icy crust have been pulled apart, creating a linear feature, a heavily contoured ridge, with a crisscross of other smaller and older ones. The hollows in the terrain are filled with darker material.

The image

This image was made from a distance of just 1,250 kilometers, and covers an area just 10 square kilometers so that it can show considerable detail. Only the Galileo spacecraft, launched in 1989 to visit Jupiter and its moons, has been so close, and made the image in December 1997.

Credit: NASA/JPL – Caltech

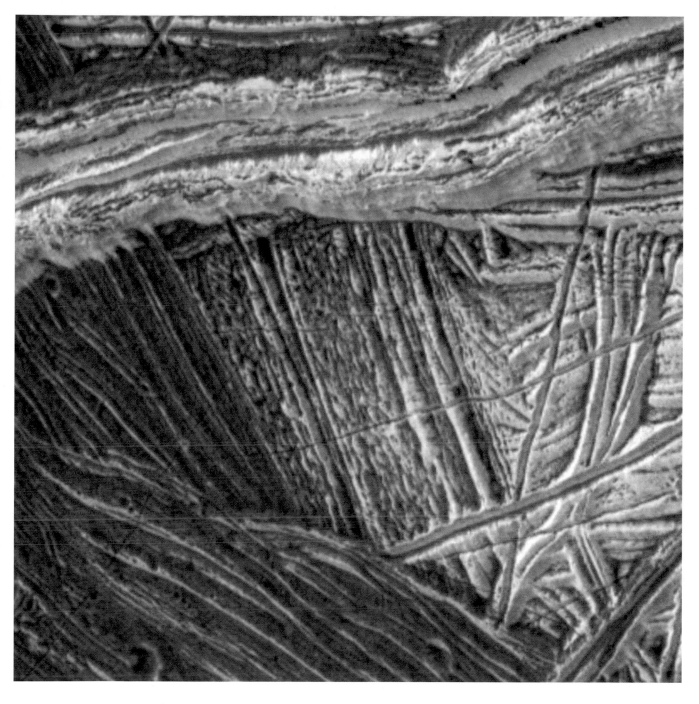

Mimas in Blue

The surface of Saturn is generally cloudy, murky, and bland. But, as in this image, the gases of the far north of the planet can have the same effect on sunlight as does the Earth's much thinner and more fragile atmosphere. They scatter blue light to our eyes, or in this case to a camera on board the Cassini spacecraft. Onto these deep blue skies are cast the shadows of Saturn's rings, while over and around the planet speeds Mimas, one of the large family of moons.

The mere presence of Mimas, as for any such marvel, raises questions.

Why is it there? It is there for no purpose, or at least for no purpose that we know of. It is just there.

How is it there? Now that is a different matter. Through careful observation, through hypothesis and more observation, we can construct coherent stories of the whole Solar System in which we, too, are contained.

Planets, moons, and all of the other objects that orbit the Sun carry messages from the past, imprinted on their surfaces and in their pathways. The messages are silent on why we, and all the rest of the wondrous complexity, are here. They don't tell us whether there is a reason at all, and we must find other ways to ponder that or just not ponder it at all. But they provide powerful accounts of the long history of our neighborhood in the greater vastness of space.

The object

Mimas

Mimas has a diameter of 400 kilometers and a mean surface temperature of −200 °C. Its density is low, suggesting that the moon is a ball made mostly of water ice – like a large snowball.

The image

This image was captured by the "narrow angle camera" on the Cassini spacecraft on January 18, 2005, at a distance of roughly 1.4 million kilometers from Saturn, and well over 1,300 million kilometers from Earth.

Credit: NASA/JPL/Space Science Institute

Taking What You Can Get

The picture is grainy. It could be an enlarged photo of distant airplane vapor trails, or, slightly more fancifully, a negative image of telephone wires in a snowy sky. It could be quite an old picture, and indeed it is, having been taken on January 23, 1986.

It is a Voyager 2 image of the rings of Uranus. It was never part of the Voyager's plans to image these rings, since they were only discovered in 1977, the same year the spacecraft set off on its journey. When it reached the seventh planet, there indeed they were. Planetary rings are, it seems, quite a common phenomenon, and all of the Solar System's gassy giants have them. (These slender bands, however, are not much like the broad disks of Saturn.)

Opportunities for taking such a picture do not come around every day. By and large they come around once, ever. And in that case you take what you can get. If it's a bit grainy then you have to work with it.

The object

The rings of Uranus

The rings of Uranus are finer and darker than those of Jupiter, Saturn, or Neptune. They are made of ice balls of different sizes, just as are the rings of Saturn, but events have covered them with a layer of dust.

The image

Taken from just over a million kilometers from Uranus, when the Voyager 2 spacecraft was flying in toward the planet, it shows objects as small as 10 kilometers across.

Credit: NASA/JPL-Caltech

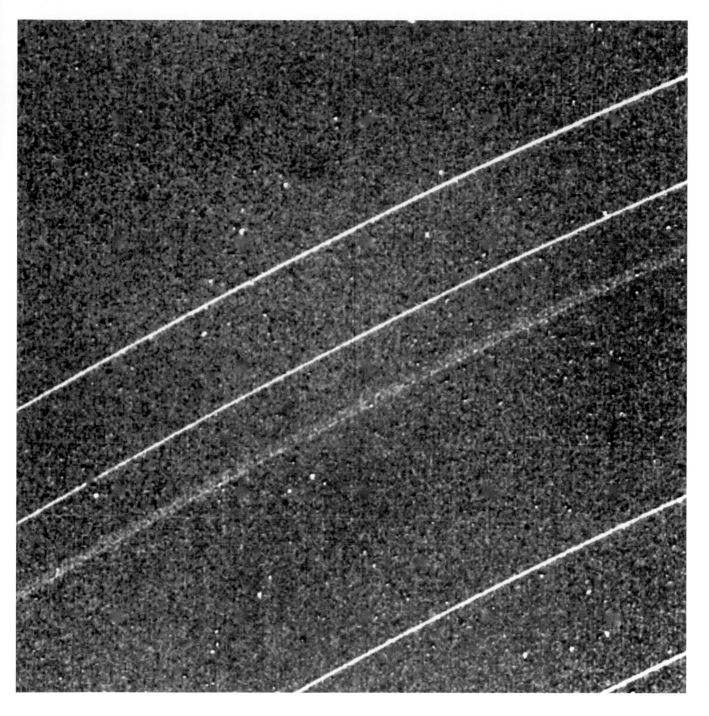

The Sun

The Earth

Comets

Time in the Sun

Not many people get to see Halley's comet more than once, since we each have too little time in the light of the Sun. If you weren't around in 1986, or just didn't look carefully enough, then you'll have to wait until 2061–2062. It visits every 76 years.

In truth, the comet didn't make a great show in 1986. It was much brighter way back in 1910, and in 1066 when it was said to predict great events (whereupon an army from France invaded England and set up a dynasty that links continuously to the present). Chinese observers recorded a visit 3,000 years ago, and every visit since 240 BC has been noted. But it wasn't until the visit of 1682, as far as we know, that anybody noted the 76-year cycle and successfully predicted the next sighting, which duly followed in 1758.

What did happen in 1986, though, was that Halley's comet was a very much observed object. To an observer getting as close as the Giotto spacecraft did, the show was not disappointing at all.

A comet seen from Earth is really an enormous stream of dust liberated by the solar wind and reflecting sunlight to us, with a much smaller solid nucleus at the center. Halley's core of rock and ice is just 16 kilometers across.

At the present rate of losing material, more than 3 tons every second during Giotto's close observation, it will not last for ever – perhaps another 100,000 years. Other comets crash into the Sun, or, less commonly, into the larger planets (and perhaps, even less commonly again, into the smaller ones). No comet that joins us in the inner Solar System has long to live, relative to the age of the Solar System itself. Just like us.

The object

Halley's comet

The Giotto spacecraft's measurements supported the idea that Halley's comet formed 4.5 billion years ago, at the same time as the Sun and the rest of the Solar System, out of ice that condensed around dust. For most of those billions of years, it must have remained with little change in the outer Solar System, until some gravitational disturbance threw it into its long cigar-shaped orbit, and toward its eventual destruction. The dust includes minerals and organic chemicals, the foundations of life.

The image

The image was produced by the Giotto spacecraft in March 1985, when the comet was still falling sunward. The spacecraft was at its closest approach to the comet nucleus, just 600 kilometers away, and deep inside its huge dust cloud. The spacecraft is named after the medieval painter Giotto di Bondone, who featured the comet as the Star of Bethlehem in his Adoration of the Magi, a fresco painted in 1304 in the Scrovegni Chapel, Padua, Italy.

Credit and copyright: ESA

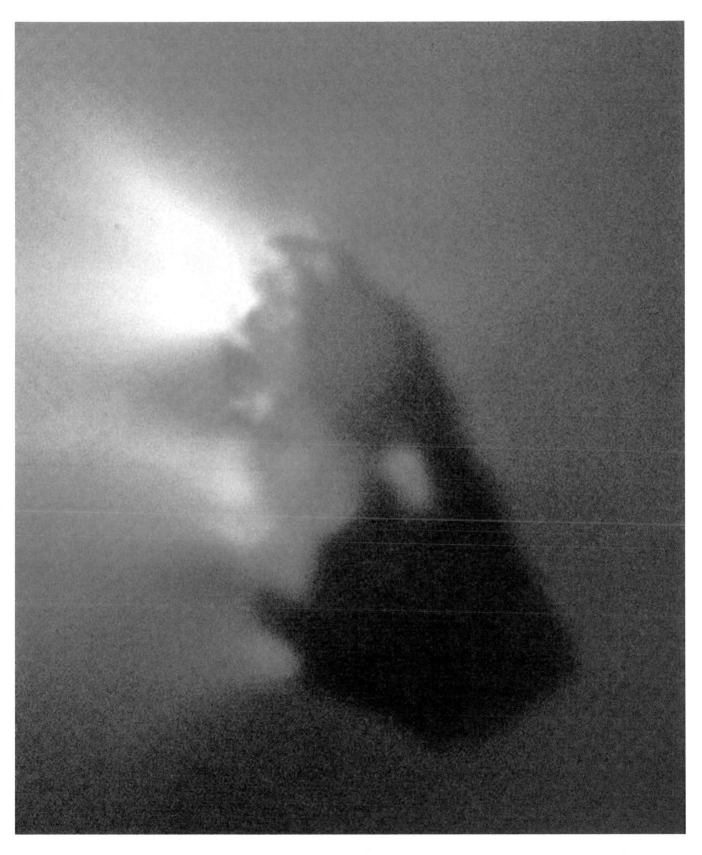

Something out there Moving

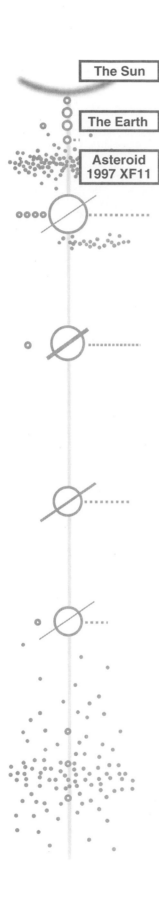

The Sun

The Earth

Asteroid 1997 XF11

Some things are easy to see, and some things are not. Asteroids, for the most part, fall into the second category. But sometimes they give themselves away by their movement.

It is not true to say that the stars do not move; the stars visible to unaided eyes are all members of our own Galaxy, the Milky Way, which cartwheels through the void. But we move with them, and any relative motion is not enough to change the visible patterns of these stars, the constellations that we see, neither over the lifetime of a single generation nor of a single civilization. Effectively, the stars form a fixed background against which we can see more local events.

So, on March 11, 1997, Bernadette Rodgers recorded six images over a 30 minute period, and the superimposed pictures do indeed show an array of fixed stars. The brightest ones have vertical streaks, a consequence of overexposure. But there is another linear track, which must be the result of something different. It is sunlight reflected from an asteroid – light about a million times too faint for direct human vision. The asteroid, 1997 XF11, was not part of a fixed background; it is very much closer than the stars and both we and the asteroid are spinning around the Sun. That very motion will bring us close together, and early assessments suggested that 1997 XF11 could be an incoming projectile with explosive potential. Or perhaps a sort of memorial stone with all our names already carved on it. Fortunately, those reports of humanity's imminent demise were premature.

The object

Asteroid 1997 XF11

This is a relatively nearby asteroid, a kilometer or two in diameter, whose path can cross that of the Earth's. Measurements of its pathway allow a reliable prediction of its future movements for some time to come. On Thursday, October 26, 2028, it will be very near indeed. When it was first discovered it was thought to be a real danger, coming close enough to the Earth to make a collision a possibility. But further measurements and computations brought some relief – it will pass more than twice as far from us as is the Moon, which is still pretty close, but life on Earth seems likely to be still here that Thursday evening.

The image

The telescope, 3.5 meters wide for effective light capture, was in New Mexico but the operator was in Seattle. The telescope scanned across the sky to compensate for the Earth's spin and to produce clear and circular images of the stars, and the presence of a noncircular feature (at the image center) tells of something in the nearer distance that is moving.

Credit: The University of Washington and The Astrophysical Research Consortium (Bernadette Rodgers, Eric Deutsch)

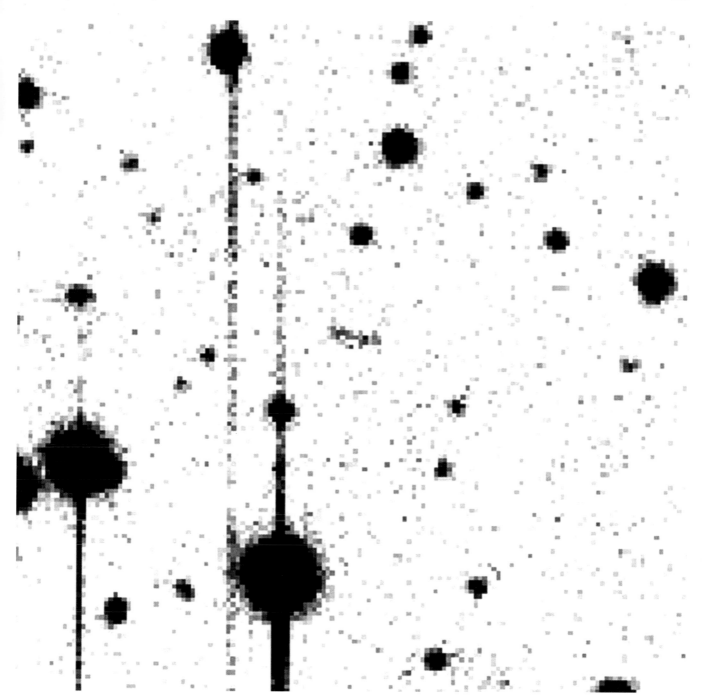

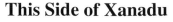

This Side of Xanadu

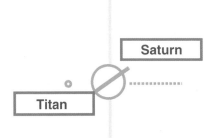

We have continents and we have seas. We have a thin layer of gas to separate us from the dark vacuum above. So does Titan.

We have people, 6 billion and more, each enveloped in his or her own unique importance. Titan has none.

It has the now-dead Huygens Lander which parachuted to its surface in January 2005. It has land and lakes. It has its atmosphere and the sounds of wind and waves and trickling streams, but nobody to hear them.

It has mists that until now have wrapped it in ambiguity. It has change, dynamism, and complexity. It has a bright continent-sized area, which from our distant continents we have called Xanadu.

But what importance does it have? That might depend on whether we, who do not hesitate to endow importance upon ourselves, are indeed the center of all creation, somehow the chosen people; or whether we are the beautiful product, one beautiful product, of a complexity that we have only just begun to recognize through the mists of ignorance. Titan, known in detail only very recently, has no obvious purpose for us. But it has profound meaning, through its existence alongside our own.

The object

Titan

Though we can see Titan's sphere, its surface is not visible from Earth. That is partly because Titan is so very far away, and partly because the smog of its atmosphere obscures its surface. But the Cassini Orbiter has peered down through the mists and the Huygens Lander has made its suicide mission, shooting images as it fell to the unknown surface. The new views have presented scientists with puzzles to solve, such as the nature of the "red spot" feature close to the top of this image.

The image

The image was made by the Cassini Orbiter in April 2005, and shows an area that is about 500 kilometers across. The relative brightnesses of the surface are real enough, but the colors are "false." The spacecraft made three separate images at three different wavelengths of infrared "light," and these have been combined together to make the moon's features more accessible to the human eye.

Credit: NASA/JPL/University of Arizona

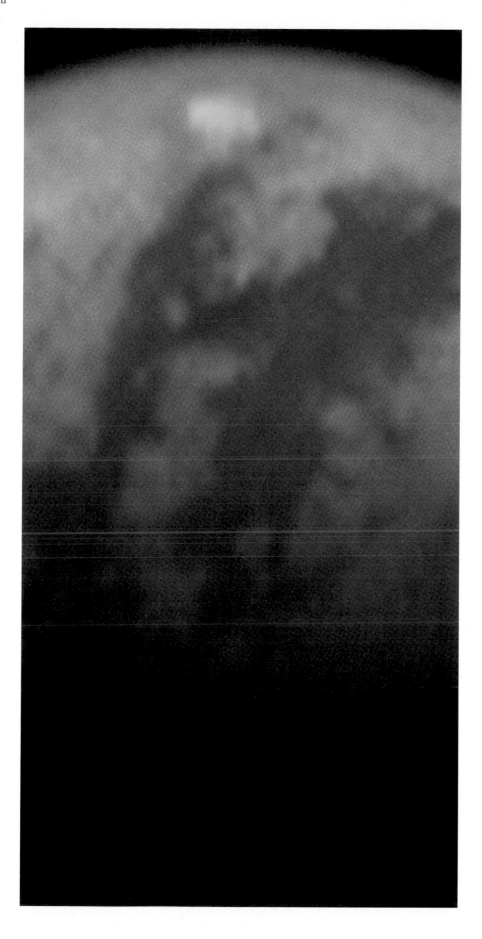

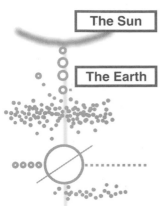

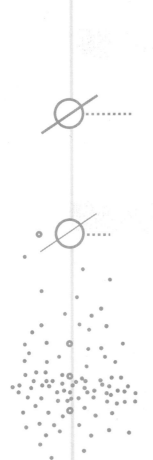

Simplicity Too

There seems to be something electric about these simple lines, like streaking car lights or a fluorescent glow in some dark place. Some might say that they seem to have human rather than natural origins. But that would be to say that humans are not natural, not part of nature, which is an arrogant and unsustainable claim. Human involvement with the glowing lines begins with the imaging processes of the Cassini spacecraft – that's a profound involvement, and a natural one.

The source of the light is the Sun. The objects scattering the sunlight to the camera are the multitude of icy grains of the rings of Saturn. Saturn itself is on the right, and the further rings (the upper rings in this image) gradually fade behind the planets absorbing upper atmosphere.

This view of the rings is simple. Nature is so complex that it can do simplicity too, without effort.

The object

The Rings of Saturn

Saturn's most prominent rings are given unpretentious names. Working outward from the planet, they are C, B, and A, and these are the structures that are visible in this image. The D ring is closest to the planet, but too faint to be seen here. Outside the A ring, the neatness of the naming system breaks down, due to the chronology of discovery, and, moving on outward, first comes the narrow but interesting F ring, followed by G, and then E.

The image

The picture shows the dark side of Saturn, with sunlight being scattered by the rings. At this low angle of view, the gaps between the various circles are invisible, and the dark band separating each pair of stripes is a dense ring through which the sunlight, shining from behind and beyond the rings, cannot penetrate.

Where light from the further side of the rings passes through the upper atmosphere of Saturn, it carries information to the spacecraft about Saturn's outer gases.

The image was made by the Cassini spacecraft in March 2005.

Credit: NASA/JPL/Space Science Institute

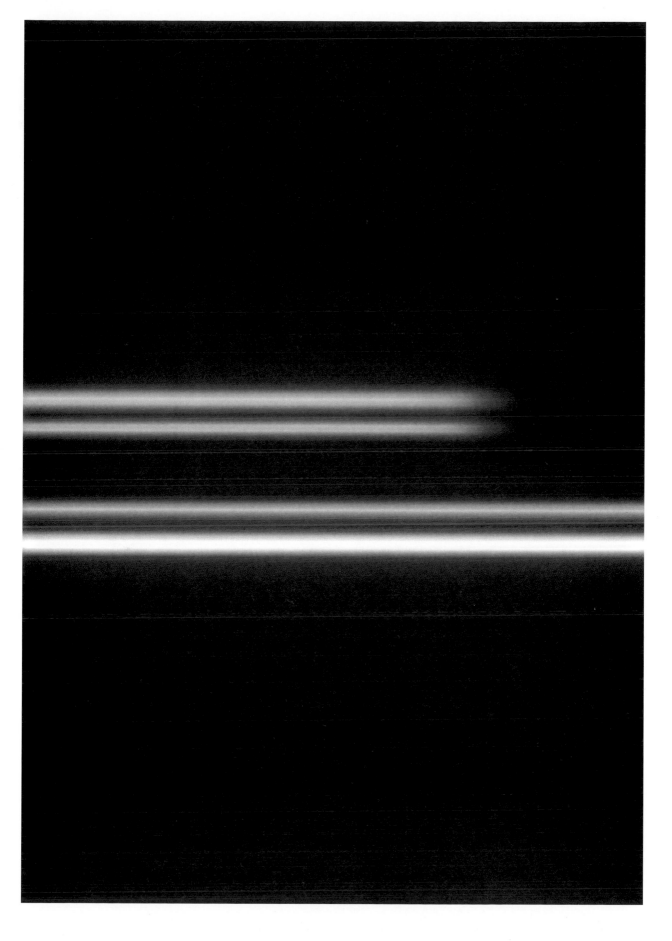

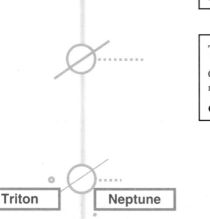

The Journey of a Hero

The Voyager 2 space mission was launched in 1977, and the spacecraft is now far beyond the orbit of Neptune traveling on outward at a speed of about 60,000 kilometers per hour. At that rate, it will reach the outermost extremes of the Oort Cloud in about 20,000 years, and will be at the distance of the nearest star after roughly 75,000 years. Radio signals now take 11 hours to reach us from the little probe, but reach us they do. It is still sending information, using a transmitter with a power of about 25 watt. To know what it has seen, we need sensitive receivers as well as an 11 hours wait. Voyager is on an epic journey of a hero.

As it passed by Neptune in 1989 it took in a quick visit to Triton, the planet's largest moon.

The object

Triton

The picture shows a few craters, which suggest that Triton has an active surface. Indeed, this suggestion is supported by the unexpected discovery of active "ice volcanoes" on the moon. At a surface temperature of −235 °C (not quite as cold as anything can ever be, but close), any kind of volcano seems unlikely, but there they are – a 140 kilometers plume of material is visible at the top right of this image.

Triton is a doomed moon – energy leaks away from its motion due to tidal interaction with Neptune itself, and it edges closer to the planet. Sooner or later, probably very much later, it will break up under Neptune's pull and form a ring, or it will simply crash down. For the time being it is a very cold object, with a bright, reflective surface of ice – solid water, carbon dioxide, nitrogen, and methane. (Earth and Titan are the only other bodies in the Solar System with significant concentrations of nitrogen in their surface layers.)

The image

Given the distance that Triton is from Earth, the detail of this image is impressive. Only the Voyager missions could have produced it. This image dates from 1989.

Credit: NASA/JPL – Caltech

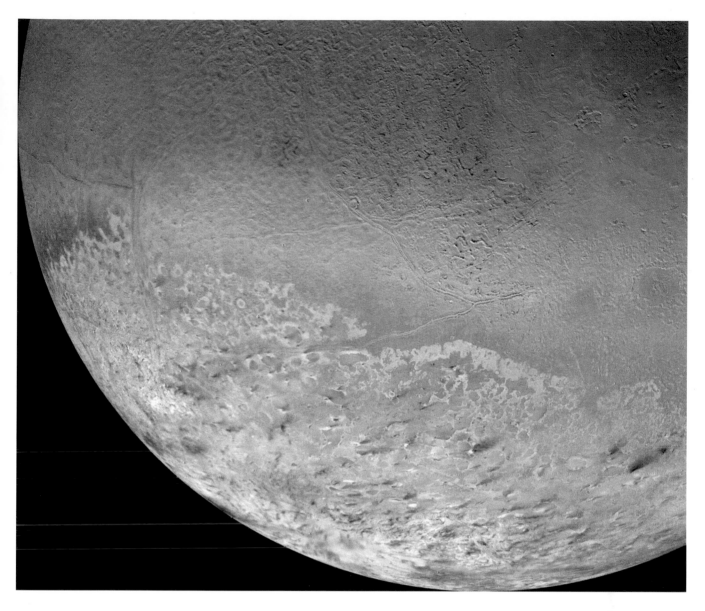

Whatever Seasons

The Hubble Space Telescope, hanging in orbit above the Earth's distorting atmosphere, is one the great success stories of space exploration. It has produced stunning new images of hordes of galaxies at the furthest limits of visible space and images of objects in the Sun's back yard. Pluto is one such object, but even Hubble could not see details on Pluto's surface, only general patterns of dark and light.

Here Pluto's globe is projected onto a rectangle, profoundly distorting the relative sizes of areas but usefully revealing lighter zones at north and south toward the poles, and patchier areas across a broad central strip.

If there were a Hubble Space Telescope in orbit around Pluto, and looking inward toward the Earth, then if it could see us at all in the blaze of the Sun, it might detect similar patterns, distributions of dark and light with seasonal variations. The similarities, though, are not likely to go much further. Pluto is significantly smaller than the Earth, smaller than the Moon, and it is between 30 and 50 times further from the warming Sun. Whatever seasons Pluto has, they are all extremely cold.

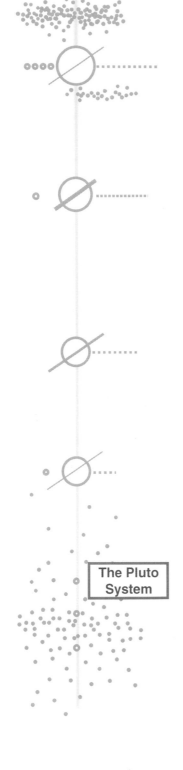

The Sun

The Earth

The Pluto System

The object

Pluto

Pluto is the size of a largish moon, smaller than people guessed when it was first discovered and declared to be a planet. It is, perhaps, just a moon without a mother planet, or a nearer member of a large number of similar objects in the Kuiper belt beyond the orbit of Neptune. Its orbital orientation, outside the plane of the paths of all the other planets, adds support to the notion that Pluto is not part of the Sun's initial family of planets.

Though it is quite small, being so far away from the heat of the Sun and with a surface temperature of just −228 °C, Pluto is able to hold on to an atmosphere. It reflects enough light to allow spectral analysis, and nitrogen, methane, carbon monoxide, and ethane have all been identified. The weak sunlight is likely to be enough to create temperature differences that drive surface winds.

The image

The image was produced by the Hubble Space Telescope in 1994. Hubble was able to observe the rotating planet over a period of about six and a half Earth-days, this being equivalent to one Pluto spin, one Pluto day.

Credit: Alan Stern (Southwest Research Institute), Marc Buie (Lowell Observatory), NASA, ESA

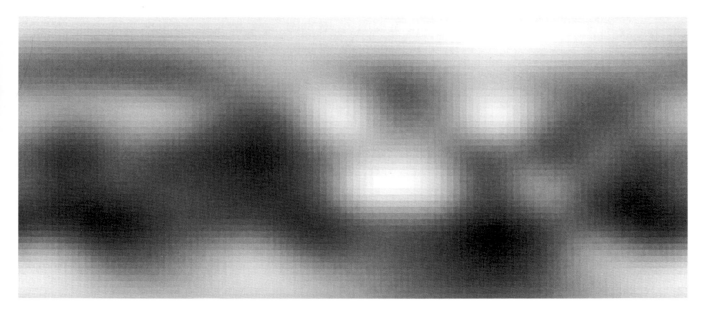

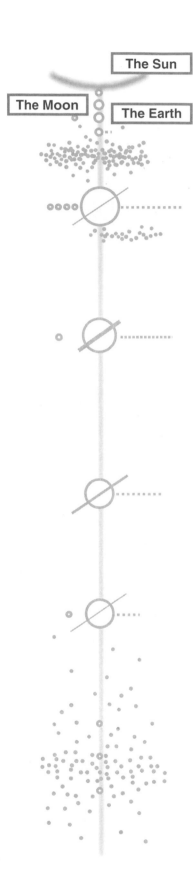

Forever Entangled

They float like a mother whale and her calf. They are indeed bound tightly together, and also to the Sun which here illuminates their hemispheres. It is gravity that binds them and determines each one of their ever-repeating motions.

They are not parent and offspring, and are very different, not just in size but in inner structure and its outward manifestations such as magnetic field, and in the nature of their surface activity.

One has an ancient surface, tickled by space dust and occasionally dented by something larger. The other has an outer layer in perpetual change; change that is sometimes deep and sometimes shallow, explosively fast at times, but always and endlessly grinding from one millennium to the next. Waves and tides flicker quickly across its oceans, while the land itself heaves and jolts and oozes, slowly making mountains that wind and water wear away.

One has no atmosphere, and is exposed to the empty horrors of space – to the vacuum, the radiation, and the blasting rain of dust. In the skies of the other, there are skimming, spinning patterns of white and blue, as water makes its subtle transformations between liquid and gas. Over, on, and under its surface there proceeds an interactivity that we struggle to understand both because of its complexity and because we are part of it, forever entangled.

Earth and Moon, a relatively small planet and a relatively large moon, are probably both products of the same event. Neither is older than the other; both are ancient, and forever entangled.

The objects

The Earth

The Earth is a planet and not a "smaller body" of the Solar System, but surely worthy of special mention as our home, whoever we may be.

The Moon

Illuminated and heated by the Sun on one side, and dark and cold on the other, the Moon has no protective atmosphere. It shows no sign of inner activity, either – no surface movement or external magnetic field. It is very nearly an inert ball of rock – bombardment of its unprotected surface by Solar System debris produces the only activity.

The image

The image has been put together from two images to make a point – about the relative sizes of Earth and Moon. The two bodies are shown to the same scale, but the distance between them is roughly 30 times bigger than the diameter of the Earth itself and here it is foreshortened. The images were made by the Mariner 10 spacecraft, on its way to investigate Venus and Mercury, in 1973.

Credit: Northwestern University, JPL, NASA

Index

Names that occur with high frequency throughout the book are not included in the index. These names are: